W0079776

Palgrave Studies in Sound

Series Editor
Mark Grimshaw-Aagaard
Aalborg University
Aalborg, Denmark

Palgrave Studies in Sound is an interdisciplinary series devoted to the topic of sound with each volume framing and focusing on sound as it is conceptualized in a specific context or field. In its broad reach, Studies in Sound aims to illuminate not only the diversity and complexity of our understanding and experience of sound but also the myriad ways in which sound is conceptualized and utilized in diverse domains. The series is edited by Mark Grimshaw-Aagaard, The Obel Professor of Music at Aalborg University, and is curated by members of the university's Music and Sound Knowledge Group.

Editorial Board
Mark Grimshaw-Aagaard (series editor)
Martin Knakkergaard
Mads Walther-Hansen
Kristine Ringsager

Editorial Committee
Michael Bull
Barry Truax
Trevor Cox
Karen Collins

More information about this series at
http://www.palgrave.com/gp/series/15081

Seán Street

Sound at the Edge of Perception

The Aural Minutiae of Sand and other Worldly
Murmurings

palgrave
macmillan

Seán Street
Faculty of Media and Communication
Bournemouth University
Poole, UK

Palgrave Studies in Sound
ISBN 978-981-13-1612-8   ISBN 978-981-13-1613-5   (eBook)
https://doi.org/10.1007/978-981-13-1613-5

Library of Congress Control Number: 2018948170

© The Editor(s) (if applicable) and The Author(s) 2019
This work is subject to copyright. All rights are solely and exclusively licensed by the Publisher, whether the whole or part of the material is concerned, specifically the rights of translation, reprinting, reuse of illustrations, recitation, broadcasting, reproduction on microfilms or in any other physical way, and transmission or information storage and retrieval, electronic adaptation, computer software, or by similar or dissimilar methodology now known or hereafter developed.
The use of general descriptive names, registered names, trademarks, service marks, etc. in this publication does not imply, even in the absence of a specific statement, that such names are exempt from the relevant protective laws and regulations and therefore free for general use.
The publisher, the authors and the editors are safe to assume that the advice and information in this book are believed to be true and accurate at the date of publication. Neither the publisher nor the authors or the editors give a warranty, express or implied, with respect to the material contained herein or for any errors or omissions that may have been made. The publisher remains neutral with regard to jurisdictional claims in published maps and institutional affiliations.

Cover illustration: © Melisa Hasan

This Palgrave Pivot imprint is published by the registered company Springer Nature Singapore Pte Ltd.
The registered company address is: 152 Beach Road, #21-01/04 Gateway East, Singapore 189721, Singapore

'Tis the deep music of the rolling world...
Percy Bysshe Shelley: *Prometheus Unbound, Act 4*

To Jo

PREFACE

This book is about the interrelationships between hearing and listening, looking and seeing. In both cases, one does not presuppose the other; we are perfectly capable of hearing without listening, just as seeing takes us beyond the utility of simply looking. Likewise, it is the hope behind this book that by awakening the faculty of seeing, we may enhance our ability to listen—and vice versa. Hence, there is quite a lot of visual art in this book, but I hope the reader will join me in the galleries I visit on the way, because it is not as an art critic but as a seeker after sound that I make these excursions. In my other writings, radio—in all its changing times and formats –has been a key element, and it is a thread here too, for after all, does not radio contain the most memorable pictures of all? It is also, as I never tire of discussing, a poetic medium in essence, and the argument of the book utilises some of the word/sound images and pictures that poetry provides in its gallery of the imagination. There is then, here a kind of tapestry of art, poetry and a sense and awareness of Place, triggering sound by association, memory and empathy with our surroundings, our circumstances and our fellow inhabitants of the planet, heard and unheard. It is about the murmurs of voice and locality, the whispers in the air, the water and the landscape, the brush of sand over rock, the faint single notes that build from a fragile hint of a presence to become part of the symphony of sounds with which we are surrounded, the sonic context of life.

These are the component parts of our sound kaleidoscope, and my preoccupation here is with deconstructing processes culminating in the homogenised end-product of a choir of voices that sometimes makes music, sometimes noise. It all begins as an instant of sound, whole and complete in itself, as a single drop of rain makes its mark on stone, before the deluge drowns it and it becomes engulfed in the flood. As before, particularly in my last work for Palgrave, *Sound Poetics: Interaction and Personal Identity* (2017), the narrative will be told from the perspective of a poet and a radio practitioner; the book is based principally on my own reflections drawn from a life of gathering and listening to and for sound, and attempting to translate the world into words in my poetry, a life I've been privileged to share with gifted friends and colleagues from the world of sound and literature, some of whom have been kind enough to share their insights here. This is an exploration of the significance that lies in the often-missed sounds of life and an attempt to champion attentiveness to them. It is about deconstructing the world's orchestra and hearing the subtleties that lie almost concealed in the sound work of which we are a part. To return to the analogy of fine art, I would have us consider the journey upon which we are about to embark as an examination of the act of making. The *pointeliste* technique of artists such as Seurat may be a clue to uncover meaning, purpose and intent within this context: tiny dots of light and colour that go to make up, through their presence and company, the whole picture. We will begin then with what we might well call 'talking pictures.' Sometimes these pictures are on walls, sometimes they are moving around us and sometimes they are within us, but always they speak, and what they have to say is key to our understanding of the world, if we have eyes to hear and ears to see.

Liverpool, UK   Seán Street
Spring, 2018

Acknowledgements

I am indebted to the following for their kindness in sharing thoughts, expertise and insights: Chris Brookes, Edwin Brys, Pat Collins, Alan Hall, Jeremy Hooker, Julian May, Eleanor McDowall, Piers Plowright, Cheryl Tipp, Chris Watson and Jennifer Wong. I am grateful to George Hyde and Malgorzata Sady for permission to quote material from their translation of *A Poem About Lublin* by Józef Czechowicz. The poet Katrina Porteous offered valuable reflections on her location-based poetry, and generously allowed me to quote from unpublished writing, as well as directing me to some of her published work on the subject. My thanks to Grażyna and Marcin Stachyra for opening my eyes and ears to the implications of certain sounds of Place. Emma Smith was generous too in sharing her thoughts on the social life of human sound, in the context of her 2018 exhibition, *Euphonia*. My thanks also go to Mark Grimshaw-Aagaard of Aalborg University and Joshua Pitt of Palgrave for their faith in the project, and as always to my wife Jo for her expert eye, patience, advice and encouragement.

CONTENTS

Introduction: The Bell of Józef Czechowicz—The Importance of Minute Sound Moments

Abstract The introductory chapter sets out the main themes of the book, opening with a discussion of sounds heard but not listened to, and progresses through the imaginative sound generated in the mind by works of art. There is an examination of the relationship between seeing and listening, using visual art as a means and point of focus. A crucial difference is the relationship between time and space in visual images and sound. Sound travels through time. Minute sounds when placed in the context of a place can act as mnemonics, awakening an array of recollections, and trigger to the imagination, assuming significance far in excess of their actual presence, acting as portals opening hidden worlds of thought.

Keywords Art · Sound · Radio · Memory · Imagination

PICTURES AT AN EXHIBITION

I am sitting in the cafe of the Walker Art Gallery in Liverpool. It is a Tuesday morning, and the marble and tile surfaces are reverberating with the chatter of visitors across a wide public space full of tables: chairs scrape on the stone floor, and in the corner, a coffee machine clears its throat noisily. Standing on William Brown Street, in the heart of the city, next door is the great Liverpool Central Library, and a few steps down the road is the World Museum, the delight of local schoolchildren, while

© The Author(s) 2019
S. Street, *Sound at the Edge of Perception*, Palgrave Studies in Sound,
https://doi.org/10.1007/978-981-13-1613-5_1

just across the road is the giant St George's Hall and nearby the Empire Theatre. The whole area is a hub, and a thriving, pulsing one at that.

For now though, I sit, drink coffee and listen. At first, the place is full of amorphous noise, and the sounds merge in my head. Gradually, I focus and begin to hear some perspective and direction—voices murmuring on the far side of the room, to my left, a sudden burst of chatter and laughter as a school party comes in out of the rain. They pass through into a teaching room, and the sound subdues somewhat. Someone laughs nearby, to my right, and then I become aware of something I had not noticed before, a low humming sound that I cannot quite identify. Air conditioning? Perhaps. Now it has gone. In fact, I realise as it stops that it is its absence that has drawn my attention to its previous presence. The voices around me have quietened a little. Is it because, without realising it, everyone has adjusted to the fact that now there is less noise to talk over, for which to compensate? How much quieter we would be if our voices did not have to compete with so much of the random cacophony that surrounds us, not least the ubiquitous soundtrack to eating so many restaurants are determined to inflict upon us. Mercifully, there is no musical accompaniment to coffee here, but there is—now I listen again—something else—a low rumble, less a sound in fact than a vibration. It comes, lasts about ten seconds and fades. Minutes pass, and there it is again. Then, I remember that in this part of the city, the railway goes underground, and these are local trains travelling the subterranean loop beneath me. Anyone familiar with the pulsing beats of night clubs will know the feeling of music that is as physical as it is auditory: rhythm will play a major part in our story, and the vibrations of low-frequency bass notes fostered by high output sound systems are palpable, and the experience is a whole-body one. If I were to record, amplify and analyse what I'm feeling beneath my feet, and slow it down, I would begin to dissect the elements of the sensation into separate vibrations, and the high volume, low throb of club dance music makes for an even better example: 'Bass notes in the ground, underfoot, in our bones. Slow the sound down further and each vibration might be separated out, counted, added up; there is no more sound, just individual shocks, one at a time' (Trower, 1).

Emerging from this imaginative troglodyte world, I listen to the voices around me again. Fragments of conversation step forward for a moment and then retreat. More, it is about the sound of the voices: mostly women as it happens, mostly of an older generation. A mix of tones and character as one would expect, and a mix too of implied origins. Some distinctive accents, clearly local voices, but there is a Spanish couple sitting not far away, looking at a map, and I can here American voices, east coast I would guess, animatedly discuss a Rembrandt self-portrait they have seen upstairs. I cast my personal audio antenna elsewhere, scan the area for its stories in the realisation that noise, once it comes into focus, and once we drill down through its layers, becomes particularised sound.

I place my cup down rather noisily. If the room were empty and silent, that sound would have bounced around and reverberated, but as it is, the sound of it is lost in seconds and no one else notices. I go back to absorbing the incoming waves of sound from around me: near sound, far sound, sound in the middle distance, a miniature indoor landscape but in sound—a soundscape. Here, in the middle, at the point of processing—the control centre within my head—this hubbub is quickly sorted, unravelled, separated, analysed and interrogated. Some of it is thrown away—at least consciously—some is recycled, while some becomes the subject of mystery and conjecture in the mind. That is probably the material I will retain, the information my imagination and memory can use as part of today's life experience. Often, it is the effort of locating the sound under the surface—under the layers of other sounds—that can prove to be revelatory. The composer and musician Pauline Oliveros expressed it well:

> Listening for what has not yet sounded – like a fisherman waiting for a nibble or a bite.
> Pull the sound out of the air like a fisherman catching a fish, sensing its size and energy...
> There are sounds in the air like sounds in the water.
> When the water is clear you might see the fish.
> When the air is clear, you might hear the sounds. (Oliveros, 50)

This book is about the tiny sounds of the world and acquiring the ability to listen to them. Often they lie buried under the utility of modern life and so we miss them, but of course they are themselves a part of that raucous orchestra, a contributing factor in the composite sound-picture of the world, clues to where we are and why we respond to life as we do. I have worked with sound all my life: I have made radio, I have taught and researched it, and I have written poems about it. This then will be a poetic and philosophical journey rather than a scientific one. Here, I will be concerned with the emotional effect of these moments of sound and our capacity and potential to develop deeper responses to them as micro-events. Thus, in some ways, this writing is a sequel to my book *Sound Poetics*, in which I explored the nature of sound as part of personal identity and interaction. Sound is essentially poetic; it shares the capacity of the poem of which it is so key a component, to stimulate images in the mind that make every individual sound experience unique. Even when little else is left, a barely heard sound is sufficient to unlock a raft of associative thoughts and ideas. A very small sound, given knowledge and the context of its history, becomes hugely significant, just as the sound of a coffee cup placed in a saucer within a large, reverberant empty room can dominate the space in which it sounds. Yet even when the room is full and noisy, the sound is still there; listening by hearing—like seeing by looking—can be a learnt skill, the sweet art of noticing things as the circumstances around us dance. It can be found intuitively or consciously, but we should nurture it carefully in the everyday. There is never a moment in our lives when it cannot enhance our experience. A tiny sound can be as eloquent as a voice whispering to us. Indeed a phoneme is the start of all stories—a breath, a sigh, a laugh or a cry; these utterances need no language to communicate their meaning. By speaking a name, we give a person to the world. Were we able to tune ourselves to the subtleties of the natural world, we might share the super-sensitivity of members of the bird and animal kingdom to sense the message in the apparently innocuous fragments of sound around us that trigger realisation of a coming storm, earthquake or tsunami.

Here with me in the cafe, I turn to the book I have been reading. Although I am in the heart of a great city, this book, a novel by Italo Calvino, *The Baron in the Trees*, surrounds me with arboreal whispering and running sap. Calvino wrote about a young Italian nobleman of the eighteenth century, Cosimo, who rebels against parental authority by climbing into the trees, where he remains for the rest of his life. (It occurs to me that this gallery will probably have portraits of young contemporaries of the fictional Cosimo from the Age of Enlightenment within its rooms.) During the course of his life among the branches, Cosimo discovers sonic wonders unknown to his earthbound friends and family. After all, those who spent their lives in the flat world could never understand the murmurings that came to one 'who spent the nights listening to how the wood packs with its cells the circuits inside the trunks that mark the

years, and how the patches of moulds spread in the north wind, and how
with a shudder, the birds sleeping in their nests tuck their heads in where
the wing feathers are softer, and the caterpillar wakes, and the shrike's egg
hatches' (Calvino, 96). I must seek these same powers of awareness, learn
with Cosimo and—wherever I am, in the town, by the sea or walking in
the countryside—tune my mind's radio to a new set of frequencies.

I will explore also the everyday sounds that may or may not be small,
but that familiarity has bred, if not a contempt, then perhaps exclusion
from our conscious awareness of their presence. We cannot listen to
everything, but we can at least make our listening more conscious and
deliberate. It is interesting to note how the use of a recording device and
a pair of headphones can focus the attention on the sound world. The
very act of initiating this mode of listening is an acknowledgement and
commitment to entering into a relationship with a world that is essentially sonic. As a first step towards this sonic relationship with what is
all around us, it is a technique that is useful insofar as by focusing in
this way, we no longer take for granted the signals coming to us from all
around. They begin to become strange and wonderful, as they first were
as we encountered them when we burst into the world from the womb
and the shock of air and all it brought with it made us cry out. Thus, this
book is about not only training ourselves to hear, listen and interpret the
minutiae of sound, but also making the familiar new again.

Many sound broadcasters have found that the seashore is an eloquent metaphor. It is an edgeland, a geographical pause between two
states of being: a border between two worlds and a place of setting forth
and arriving. In the opening pages of his great work for radio, *Embers*,
Samuel Beckett evokes the beach as a haunted place where Henry shows
us the ghost of his father against a strange background of dark waves.
A shoreline is where the visual and the auditory are most mutually con-
joined as a complementary sensory experience. Later in this book, we
shall explore the rich and complex sounds of the natural world and seek
to delve down—and up, into Cosimo's trees—seeking out the subtle
messages it sends us, and as part of that aspect of our exploration, we
shall visit the work of the lyrical geographical scientist, Vaughan Cornish.
In the meantime, here is Cornish on his own seashore, invoking the
capacity that we shall enhance within ourselves on this journey: the ability to be in a place fully, with all the senses working as equal partners.
A beach, as he shows us, is a good place to start, because its sound is as
near to its visual presence as anything in nature:

At low water there is a soft murmur upon the flat sands then exposed, which changes to a rhythmic boom when the waves reach the steep, shingle slope. The breaker increases in height, culminates for a moment in a cusp, and then, curling over in a scroll, descends in thunder, the clear dark water transformed into a white, foaming surge which sweeps over the rattling shingle...No other aspect of the natural scene presents so perfect a harmony of sight and sound as the waves which advance upon the shore. (Cornish, 37)

Here in the art gallery cafe, there is ebb and flow too. I notice that the weave and weft of the voice tapestry around me have changed, and I see that the Spanish couple have gone, and the Americans are walking up the stairs, back to the galleries, perhaps to take another look at Rembrandt.

Our own overall sonic experience of a gallery provides the soundscape into which are placed the focused specifics of the individual art objects, each with the potential to evoke its own sound world in the imagination: internalised sound placed within the context of physical sound. Leaving the room, we carry the memory of it, of the objects we have seen, the impression they made upon us, but also of the place itself. Physically, it may have been a shared experience with other visitors, yet in audio terms, every personal response remains unique and of our own making. As Franz Kafka said: "Everyone carries a room about inside him. This fact can even be proved by means of the sense of hearing" (Kafka, 1). We may choose to purchase a catalogue as a souvenir of our experience; when we open it in a new environment, say our home, office or classroom, the sound of this new place provides a changed audio backdrop, while the 'music' of the images as we turn the pages may—or may not— remain the same as on first seeing. A broad analogy would be the experience of *live* music in the concert hall, contrasted with a recording of the same artist and/or work listened to within the environment of the home; one provides a direct witness to the event, while carrying with it all the unpredictability such a happening brings, while the alternative of a reasonably controlled situation, listening to the same work as a purely auditory experience may aid contemplation, but through the medium of a copy. One could argue that neither is totally definitive, and indeed such a reading of the performer's original intention could only really occur were the *live* performance itself to take place in the presence of the listener alone. Nevertheless, we find ourselves returning to experience the original in situ in order to replenish our mnemonic and thereby nourish our experience of the work itself. In other words, a painting performing *live* is a happening of its own. Later, we shall explore the affect of an auditory work when augmented by the visual and acoustic properties of great spaces, such as a church, in which the sonic properties of the space itself as a part of experience are important in the event and its memory. A dry external acoustic compared to the liquid reverberations of a cathedral is a part of the performance of the world.

I make my way towards a room containing a mixture of nineteenth-century paintings, including some Impressionist work. On a landing, I pause beside a large portrait of the jazz singer, art critic, writer, lecturer and bon viveur, George Melly, who died in 2007. Melly came from Liverpool, and this picture is one of a series painted by his friend, Maggi Hambling. George was a larger-than-life personality, a life force if ever there was one, and Hambling's pulsing canvas captures that exactly, positively fizzing and vibrating with the energy that was George Melly. If ever a picture was audible, this one is, and its soundtrack is 'Dr Jazz.' As a boy, Melly's world was south Liverpool, where I live and I know the roads he knew well. In the autobiography of his childhood, *Scouse Mouse*, he writes vividly of those days between the wars and of some of his seminal experiences from the time. I remember that he bought his first sheet music, of Bing Crosby singing 'Pennies from Heaven' one day from a music shop near his home, and that he retained the memory of the 'pale blue art-deco cover and a round inset of the young Bing Crosby in the bottom right-hand corner' (Melly, 176). As he grew, the moment, the place and the song grew together, so that fifty years later he was able to write that 'for some reason I can "see" the cover of "Pennies from Heaven" superimposed, like a pop collage, on the sky above a road that led up from Aigburth Road to the bottom of Mossley Hill' (ibid.). That is on my way home; from now on, I am sure I will hear Crosby's voice and 'Pennies from Heaven' every time I travel the Aigburth Road, and I shall think of George.

I walk into a relatively small room of about twenty pictures. Although it is not a large collection, cumulatively it has the same effect on my senses as did the cafe when I first arrived—twenty-three voices calling from four walls, their community for a time obscuring their individuality. It is as though I, a stranger here, have imposed myself on a gathering, a conversation in progress. It is rather the same as entering a room full of people, say at a party. The first impression is of a crowd from which one is excluded, and then the process begins of selecting the component parts. There may be one face that draws you in, someone known or familiar perhaps, or it may be a voice or a personality to which you are attracted for reasons which may not be at first apparent.

I notice a picture towards one of the corners of the room. The picture is at first glance, not particularly dramatic. It is a beach scene on an overcast day, looking out to a dark grey and fairly featureless sea. The shore is different to Vaughan Cornish's dramatic confrontation between sea and land; here the passage between the worlds is a shallow, gradual one, with lapping waves gently insinuating their sound over the sands almost imperceptibly. There are two distant yachts and a hint of mackerel back clouds above them. Perhaps the weather is changing. The colours are subtle, muted: beiges, greys tinged with pale pink and occasional touches of orange. The beach is rather rough, there is a rubble of stones and small rocks, and the sand looks rather muddy. 'Low Tide at Trouville' by Gustave Courbet, painted in 1865. A small shock of recognition: I know this view. I have been to Trouville, a long time ago. We stayed with friends nearby, at the town's somewhat more glamorous neighbour, Deauville. I remember the sand and the contrast between the two places. My memory is that on Deauville beach, the sand was more golden, paler, softer, somehow more refined and chic, and the spring breeze (it was April) wafted and whisked it in a gently fine spray across the boardwalks between the beach huts and the rows of sentry-like folded beach umbrellas. I remember the low hiss of the sand on the wood and, beyond it, the distant 'shhh' of the sea across the wide beach.

I come back to Courbet and Trouville, and I read the caption beside the picture. From this, I learn that Courbet (1819–1877) had grown up in inland France and did not see the sea until he was 22. It was a revelation to him, changing the way he used colour. He wrote to his parents: 'We have at last seen the horizonless sea; how strange it is for a valley dweller. You feel as if you are carried away; you want to take and see the whole world.'[1] Here is his voice; I hear the young Gustave in all his exhilaration, and now I imagine him feverishly sketching this scene in the moment. Think of that: at 22, being a part of all this broad land/ seascape for the first time, hearing the constant murmur of the surf, the cry of the gulls, the buffet of the wind as the evening came on. I listen in imagination to his young voice speaking the words, and I hear the place in the time and his presence in it. Most of all, I hear in my mind the personal sound between us, sharing this encounter across time, while being all the time aware that we are surrounded by other wall-hung conversations waiting to engage with me. This though has been a special encounter, triggering memory and implication. I turn away reluctantly and glance back as I leave the gallery. I move out of range of Courbet's picture, but somehow I can still hear it.

The concept of a piece of sound art has become increasingly accepted within gallery culture. In the summer of the year 2000, the musician and writer David Toop curated *Sonic Boom: the Art of Sound* at the Hayward Gallery, London. Toop was conscious of the context in which we hear sound: the inspiration for Akio Morita and Masaru Ibuka to develop the Sony Walkman had been, after all, the ability for listeners to avoid conflicts caused by competing external noises while listening to music. John Cage's response to listening had been the opposite. In his *4'33"*, he had encouraged the audience to consider the sound of the location in which the so-called silent piece was 'performed.' Likewise, Toop, in his introduction to the Hayward Gallery exhibition, wrote:

> All the artists in Sonic Boom are alert and responsive to the richly clamorous environment in which we are now immersed. Rather than searching for ways to cancel out the murmurings, hummings, pulses, whistles, alarms, signals, irritations, pleasures and shocks of the contemporary soundscape, they focus on their essence, impact and effect, so shaping new meanings for a bewildering range of aural events. (Toop 2000, 15)

This is therefore a self-conscious acknowledgement of the sound environments within which we experience art, whether it be sonic or apparently silent, just as, in John Berger's words, 'Photography is the process of rendering observation self-conscious' (Berger 2013, 19).

It is exactly what we do within our head when we read a poem or a book, a mental process that gives us the instrumentation to orchestrate the printed codes into imagery. In fact, the internal process goes further, turns three hundred and sixty degrees, because it takes a picture, be it a visual or an audio image created by another mind, filters it through the neutral medium of words and reinvents it through personal experience and circumstance to make a drama that in turn is mitigated by our own personality and placed in our memory bank. Gilles Deleuze has written: 'Musical art has two aspects, one which is something like a dance of molecules that reveal materiality, the other is the establishment of human relationships in their sound matter' (quoted in Stenger 2014, 15). The miracle of composition is the revelation of patterns of sound placed on silence that touch a chord of recognition in us. We are each of us composers, and our orchestra is our imagination.

BRUEGHEL'S SOUNDTRACK

Can a work of visual art really contain sound? To answer this, we must move beyond physics into the imagination, even into the realms of the metaphysical. The poet and theologian, Rowan Williams wrote: 'silence and the attentiveness that silence brings imposes, opens us up to communication from sources we should otherwise ignore.'[2] In a sense, every work of art has the capacity to be 'sound art' insofar as we possess the capacity to sonically incarnate and articulate any visual image presented to us. We can verbalise our response to it of course, describe it in language or philosophise upon its content as for example, did the poet W.H. Auden on Pieter Brueghel the Elder's 'The Fall of Icarus' in his poem, 'Musée des Beaux Arts.' In such ways, we bring an object into the world and consciousness through new ways, giving a thing a life independent of, but dependent upon its original manifestation. Before this, however, the work has to 'speak' directly to our consciousness, and to recognise its sonic potential is to develop layers of meaning that can greatly enhance its richness to us and the complexity of our response. In other words, we *translate* it. To illustrate this, let us take as an example another painting by Pieter Brueghel the Elder, known variously as 'The Hunters in the Snow' and 'The Return of the Hunters.' Painted in 1565 as one of a series, five of which survive, depicting different times of the year, the original is housed in the Kunsthistorisches Museum, Vienna, and it is a useful example to discuss, partly because of its obvious sonic qualities, but also because it is one of Brueghel's most famous works, familiar to many and widely available in second-hand images.

The scene is set in the depths of a European winter, during December or January. In it, three hunters are returning with their dogs from what would appear to be an unsuccessful hunting trip. They trudge wearily through the snow, and their dogs hang their heads. One carries the body of a dead fox, an indication of the paucity of their efforts, and there are the footprints in the snow of a small animal, possibly a rabbit, showing missed opportunities. They have come to the brow of a hill, and below them is their village stretching before them towards a strangely spectacular mountain range, clearly uncharacteristic of the rest of the view, and certainly invented rather than painted from reality. It is a mythic landscape imagined on a still, overcast day, and the snow on the ground would seem to be fresh-fallen. Skaters are moving across frozen ponds, playing hockey and curling; there is a frozen waterwheel, birds swoop from the bare, leafless trees above the men's heads, and nearby several adults and a child are preparing food at a fire outside a wayside inn. It is all muted in terms of colour, and the smoke rises straight and hangs in the windless air. Absorbing the scene, we may find ourselves identifying with the cold of the scene. If, however, we allow ourselves to absorb the suggested *sound* within the painting, we move into a three-dimensional, stereo perspective that echoes the visual impact. There is indeed a soundscape playing in our heads that runs parallel with the world depicted within the picture itself—muttered voices and whines from the hunters and their dogs in the foreground, voices and the crackling fire to our left from the inn, the cries of birds above our heads. Beyond that, there is the sound that comes to us from below, rising up from the village and its skaters, borne up to us on this bluff through the still icy air. It is that moment when, surmounting the crest of a hill, the *sound* of the scene, suddenly revealed before us, opens up, and a kind of widescreen stereo impression floods into the consciousness. It is a painting that rewards meditative study, and it is no coincidence that 'The Hunters in the Snow' has featured in a number of motion pictures; Lars von Trier's film, *Melancholia* (2011), contains its image, as does Alain Tanner's *Dans la ville blanche* (1983), and it was the inspiration for Roy Andersson's 2014 film, *A Pigeon Sat on a Branch Reflecting on Existence*. Most significant of all, because of his poetic use of sound in all his work, is its presence in *The Mirror* and *Solaris* by the great Russian director Andrei Tarkovsky. Notably in *Solaris*, in which several of Brueghel's seasonal paintings are depicted on the walls of the space station, the sonic aspect of 'The Hunters in the Snow' contributes to the presence of nostalgia for the absence of earthly humanity. The camera lingers over details in the picture, and we hear the sound of its story, the trees, birds, dogs and footfalls on snow. The landscape of earth is brought into the sterile environment of the space station. It is a longing for another time and place. Brueghel himself may have been suggesting a similar sense of a lost era at when the work was created in the 1560s, a time of religious revolution in the Netherlands. There may be here a reference to an idealised past time in rural life. This year,

'The Hunters in the Snow' will once again appear on Christmas cards around the world, and—as in many of Brueghel's paintings—this suggested sound world will speak, perhaps subliminally, in parallel with its visual message. The exercise of exploring a painting in terms of its sound may be applied to an infinite range of images, and I would suggest that anyone interested in pursuing this line of enquiry selects their own examples, thus demonstrating through personal experience the imaginative music held to a greater or a lesser degree within otherwise seemingly mute visual objects.

In the summer of 2015, The National Gallery, London, staged *Hear the Painting, See the Sound*, an exhibition in which six noted musicians and sound artists generated soundscapes to accompany a painting of their choice, drawn from within the Gallery's collections. The word 'staged' is appropriate in this context; each of the paintings selected was bathed in a spotlight, surrounded by subdued twilight, and the aural accompaniment enabled the viewer to linger in front of the works, experiencing them almost as theatre. The artists and composers chosen for the project each had distinguished pedigrees in their respective fields; natural history sound recordist Chris Watson selected Akseli Gallen-Kallela's 'Lake Keitele,' Susan Philipsz chose Holbein's haunting and mysterious picture, 'The Ambassadors,' and Janet Cardiff and George Bures Miller interpreted 'St Jerome in his Study' by Antonello da Messina. The American composer Nico Muhly used the fourteenth-century 'Wilton Diptych' as his subject, Jamie xx of the electro duo, *The xx* selected Théo Van Rysselberghe's 'Coastal Scene,' and the French film composer Gabriel Yared created a score to complement 'Bathers' by Cezanne. In introducing the project, National Gallery director Nicholas Penny, interviewed by Helen Brown in *The Daily Telegraph* on 7 July 2015, stated that 'when sounds have been composed in response to a work of art, they can encourage – even compel – concentration.'[3] While this may be true, the concept of an imposed sound commentary to a work of art is problematic, in that it can intrude on and negate the viewer's personal sonic interpretation. A poem or a piece of prose fiction is an imaginatively creative partnership between its author and the reader, and the relationship between a radio producer or sound artist and the listener exists in the same kind of balance. We may justifiably extend this analogy to our response to a visual object or event: the experience of the direct encounter is personal, and it belongs to us and us alone.

Thus, the relationship between the apparently autonomous visual image and the sound that it evokes through suggestion in the mind is more complex than we might at first imagine. We have long understood that a sound can evoke a visual reverberation of its meaning, so it should come as no surprise to realise that the reverse is also true. It is the concept of time as an element in the making of both image and sonic signal, however, that we must consider if we are to gain a fuller understanding of relationships and differences. To study 'The Hunters in the Snow' is to absorb a fixed image, but the sounds of the moment captured affect the imagination in real time. A photograph is an even more precise, fixed record of a moment, in the sense that the device which recorded it responds to the will of the photographer in an instant. The novelist John Fowles, writing an introductory essay to a book of photographs by Fay Godwin, expressed 'an almost metaphysical horror before photographs, that they freeze time so, snatch their fractions of a second from it and then set them up as the ultimate reality of the thing photographed.' Fowles used as an example, a photograph of the poet Thomas Hardy standing with a bicycle in front of his house, Max Gate in Dorset, England, taken by his friend, a local parson and amateur photographer, Thomas Perkins, in the late 1890s. It is a fixed, secure image, but as Fowles asks in his essay, 'what happened five seconds before? What happened five seconds after, when the photographer took his head from beneath the black cloth and announced that the very recent present was now eternal future?' (Fowles in Godwin, x). In such an instance, imaginary sound may help provide a temporal context. The *idea* of a place—its sonic presence such as bird song, trees stirred by the breeze, the crunch of an approaching footfall on the gravel path and sounds from the house beyond—all of these can offer a soundtrack to the silent instant of the photograph. It may not be an exact recording of the moment in the way that the photograph can claim to be, but it is a kind of poetic truth, belonging exclusively to the individual viewer of the picture. Indeed, the whole process of extending the sensitivity of our sensory powers to absorb complementary signals from the world around us requires us to apply a form of poetic response to our environment. As the poet John Keats wrote to his friend, Benjamin Bailey:

> Every mental pursuit takes its reality and worth from the ardour of the pursuer – being in itself a nothing...Ethereal thing[s] may...require a greeting of the spirit to make them wholly exist – and nothings which are

made great and dignified by an ardent pursuit, which by the by stamps the burgundy mark on the bottles of our minds, insomuch as they are able to consecrate what e'er they look upon. (quoted in Motion, 218)

A POET'S VOICE

A bell is a moment that lingers. Its sound travels through time, and as it decays, it exercises our listening. Listening is part of consciousness, but in order to employ it as a skill to the full, we should make it *self* conscious. To ask ourselves to attend to the smallest sounds in the world is a challenge. We are not encouraged by our circumstances to absorb sound; rather the opposite; as Barry Truax has written, these are sophisticated skills that are in decline in us within our urbanised situation 'both because of noise exposure which causes hearing loss and physiological stress, and because of the proliferation of low information, highly redundant, and basically uninteresting sounds which do not encourage sensitive listening' (Truax, 13).

The world is cluttered; the rubbish, the detritus of living piles up around us, and the accumulation is not only of physical things. The air is clogged with wireless signals, and electromagnetic disturbances are all around us, perhaps affecting us even when we are not aware of their presence. The sonic equivalent of all this is what the poet Katrina Porteous has called 'noise-dirt':

> Noise is to sound as weeds are to flowers: out of place, out of control. Some of this is cultural: that to which we accustom ourselves is easier to ignore. But there is more to the upsetting peripheral noise than either what it signifies or where it belongs. It is something to do with music, the mysterious way that the brain interprets rhythm, pitch and tempo as emotion. The machine noises are violent, aggressive, mindless, mechanical. The specific frequency of the shrill monotone pulse makes my skin crawl as chalk does, screeching on a board. The dumper truck's gear-change is unresolved, frustrated, goes nowhere. It feels angry, belligerent, as it rumbles through my bones.
>
> This is noise-dirt, this sound out of place, loud, discordant, nerve-jangling, stress-provoking. It has become, in my lifetime, as intrusive, ubiquitous, unchallenged, and perhaps in some ways as damaging, as plastic in the ocean.[4]

Like many other people, my hearing is degenerating. I have some tinnitus, but I have learnt to listen through it, and it only really manifests itself when conditions are very still. I have lost some top frequencies; thus, many birds and insects have disappeared from my sound register, unless I switch on a hearing aid, but this is not selective—it does not have a brain to choose what it amplifies—and in noisier environments, it is a positive hindrance. I can be philosophical about this; it can be an occupational hazard in my field of work. I have a number of friends who are severely affected by deafness, and I have seen the depression and social isolation it can cause. In some case, there is said to be a link between hearing loss and the decline of cognitive powers. The preservation of what we have is important. The world of sound is not only precious and potentially beautiful, hearing it is as much a part of our situation as touch, smell and sight. Yet within our limitations, we may celebrate what sounds we receive. For every human being, no matter how good their hearing, the search for tiny sounds with the naked ear will always stop at a certain point, because we are all subject to our physiological limits. The sonic world of bats, birds, dogs and other creatures can never be ours without highly sophisticated technical support, but within those limits we have at least a duty—which is also a joy—to acknowledge what the world around us offers. The point is exactly that: acknowledgement. This book is in particular, about listening. By recognising a sound, we make it a part of us, so everything between the covers of this book is concerned with hearing and—importantly—listening. More, it is to do with realising existences through the sonic symptoms of their presences as far as we are able. We can do no more than our best to hear these voices and to try to interpret them. We owe it to the world that we do no less.

While we cannot close our ears in the way we can close our eyes, our hearing shares the ability of our visual faculties to adjust to situation and circumstance. As our eyes adjust to variations in the brightness of light, so our ears—in concert with the brain—change our response to sound levels through what are known as threshold shifts, which relate to the variations in sound that can be perceived in any given environment. As Truax explains:

> The auditory system responds to the average noise level of any environment by shifting its sensitivity...One only has to remove all extraneous noise from an environment, dim the lights, and concentrate on one's hearing to experience the gradual dropping of the hearing threshold that progressively brings minute sounds into prominence. At least 15 minutes is required to bring hearing to its most sensitive state, a duration similar to the adjustment period for a low light environment. (Truax, 13–14)

These words are crucial to the content of the following pages and to the implementation of active listening that I wish to advocate.

Our senses are no more than feeders to the brain, and it is in the mind that the vital processing occurs, the filtering and selection of what matters and what does not, the part of our primitive survival system that decides when a sound represents danger and when it is an object of curiosity or wonder. Thus, it is the brain we must educate in order to make more sense of things and to gain the very best from the places and people with which we interact on a daily basis. We know that a tiny sound, one small voice crying, can own an importance far beyond its volume. Timbre, duration, pitch and location (the character of a sound can be vastly altered according to where it is heard, e.g. in a reverberant church or across fields and woodlands, as Thoreau experienced) are all factors, but they are not governors of meaning; this, the mind decides. The smallest imaginable sound can travel around the walls of the memory for ever, if it is associated with a moment of emotion or revelation.

On a visit to the Polish city of Lublin, I spent a morning in a museum devoted to the work of a local poet, Józef Czechowicz. As a poet myself, I knew of Czechowicz, although little of his work is available in English translation. I knew that he had been a radio man, and had worked as a writer for Polish radio between the wars, and that he was a friend of the great Polish poet, Czeslaw Milosz. Intrigued, I wanted to know more, and reading in a guide book that Lublin possessed a small museum dedicated to his work, I explored further. Czechowicz was an avant-garde poet, a journalist, a playwright and a translator. He was above all a man of Lublin and wrote of the town in *A Poem About Lublin*, a radio piece that used voices in a way that anticipated in some ways *Under Milk Wood* by Dylan Thomas. His death was abrupt and tragic; working at Polish Radio in Warsaw, at the outbreak of the Second World War, he returned to his native city, believing he would be safer there. However, on 9 September 1939, in one of the first German attacks on Lublin, he was killed. He was just 36. There is a monument to him in one of the city squares, a roughly hewn, impressive and defiant work that conveys something of the vision of the man, together with a spirit of resistance.

Czechowicz was a complex person, a melancholic, a gay man at a time when to be so was dangerous, a visionary and a perfectionist. Yet in the museum, I found little that survived of him physically. Both his home in Lublin and his apartment in Warsaw had been destroyed in the war, and little remained that one could identify as intimate possessions: a few artefacts, a pen, a writing slope, a violin that belonged to his brother, little else. There was, however, one item, a small, rather insignificant brass bell, hanging on a wall. This, I was told, had once stood on Józef's desk; when he felt his mood lower, we would ring it, as a signal to lift him to a better mental place. I asked the museum's curator, Jaroslaw Cymerman, if I would be allowed to ring the bell, to make a short recording of it. The sound was minute, audible only because of the stillness within the room. It was a deeply moving moment; in the absence of so much else, here was a sound that Józef had known, a sound that had told him of possibilities and purpose. It was as near as I could possibly come to the man. It was almost like a voice.

Now when I think of Czechowicz, I have only to play my recording—no more than seven seconds—to open a wealth of memory, emotion and mental images; I can be in the room again, talking with Jaroslaw, envisaging the corridors, the little garden at the end of a passageway, the road outside and beyond, the whole town layered outwards from there. It is better than a photograph, because it moves through time as I do, as Józef did. Later, I received an email from Jaroslaw to tell me that he had, in fact, located a translation of the *Poem About Lublin*. It is, as one might expect, full of sound conveyed through words:

> In the mist you cannot hear the steps, which
> bring the Wanderer closer to his native town.
> Field paths bulge, swell into roads, and again
> they spread wide among billowing cornfields.
> The road rolls onwards. A gusty wind whistles
> in the ears of corn. Midnight is not far off, and
> someone is still drawing water from the well.
> You can hear the gantry. Still countrified here.
> Still countrified. The moon scurries through
> the clouds. The mist gets thinner. (Józef Czechowicz, trans. Malgorzata
> Sady and George Hyde[5])

While this English translation is a treasure, in that it gives us the sense of the meaning, almost a pre-echo of the opening of *Under Milk Wood* by Dylan Thomas, every writer who renders words into another language would acknowledge that the *sound* of the original is lost. It is, as Mark Glanville reminds us, a fact that 'the fricative Polish language, which lends itself so well to the conveying of meaning through sound, is almost impossible to replicate in English.'[6] At a conference about sound,

I played my recording of the bell and told my story. Afterwards, a Polish radio producer, Agnieszka Czyzewska-Jacquemet, told me she knew of a recording of the poem in Polish radio. It was not the original made in 1936, but a new production, featuring some of the same actors. The recording, although in Polish, a language of which I have no knowledge, still spoke powerfully in its own right. The words, given dimension by the grain of the voices, were moving through time, just as the bell moves through its time. I am as far as possible occupying the same air as Józef Czechowicz, and I can summon his spirit by listening to his bell in my Proustian moment.

There is another memory from the city of Lublin related to sound... but that must come later. In the meantime, let us remind ourselves that *self*-conscious listening (in the best sense of the phrase) is a choice. A useful exercise is to allow the ears to supply information more exclusively than they do normally, when playing their part with our other sensory organs. For example, to open a door or a window onto a busy street with one's eyes closed but ears attentive is to engage actively in an aural relationship with the world. In the morning, let the first impression of the day be sonic rather than visual. It becomes immediately apparent that we have a stereo picture before and around us: traffic, weather, voices and beyond that the minutiae of individual sounds that compose the overall texture of the morning. In doing so, we might well hear things otherwise unnoticed, like the hiss of bicycle tyres on a wet road, a single bird or an approaching footstep. To extend this idea, we might now open our eyes and allow them to direct our search for sound; look closely at a specific object. Does it carry a sound? Is it audible through the general soundscape? Even if it is not, we may recognise its presence, because whether it is physically audible or not, our memory and imagination can come to the aid of our brain, and supply information related to the sound it makes, a component part of the moment.

It will of course, always be easier to conduct such exercises as this in more muted environments, as in a library reading room, or a doctors' waiting room for example. In a darkened space, every glimmer of intruding light becomes significant. Salomé Voegelin noted:

> Listening in the library draws me into the minutiae of human sounds. Every hum, cough, whisper, every footstep, sneeze, paper turn, rasp and throat clearing is amplified. In sound the library becomes an awkward space of fraught physicality: full of bodies, rigid and tense, trying to be silent. (Voegelin, 12)

I will seek to draw attention to the importance of listening to these murmurs, to reawaken our consciousness of the subtle voices—human and otherwise, intended or otherwise—that go to make up the orchestral work in which we are ourselves contributing musicians, or, to use a poetic analogy, the tiny haiku lying within the epic poem which is being constantly written around us. We will explore the everyday—the mundane sounds that become strange and even threatening when heard against a backdrop of stillness as opposed to being obscured under the layers of the all-enveloping soundscape of living. The board that creaks in the dead of night, the dripping tap downstairs, small sounds that once heard in the night, will keep us from our sleep, redolent as they become with memory triggers for emotions and thoughts that make the mind race, provoking an array of emotions beyond their scale and significance in the light of day. These are the kind of sounds that can shift the mind from its purpose. John Donne in his sermon LXXX berated himself for being so easily diverted from a conversation with God as to succumb to the small sounds surrounding him at prayer:

> I throw myself downe in my Chamber, and I call in, and invite God, and his Angels thither, and when they are there, I neglect God and his Angels, for the noise of a flie, or the rattling of a Coach, for the whining of a doore; I talke on, in the same posture of praying; Eyes lifted up; knees bowed downe; as though I prayed to God; and, if God, or his Angels should aske me, when I thought last of God in that prayer, I cannot tell: Sometimes I finde that I had forgot what I was about, but when I began to forget it, I cannot tell. A memory of yesterday's pleasures, a fear of to morrows dangers, a straw under my knee, a noise in mine ear, a light in mine eye, an any thing, a nothing, a fancy, a Chimera in my braine, troubles me in my prayer. (Donne in Dunn, iix)

Everyone will be able to empathise with Donne in his battle between concentration and trivial intruding sounds. Minute sounds, when they control us rather than we them, grow out of all proportion into shouted intrusions. Better, we might think, to surround ourselves with a consistency of sound, that, like tinnitus, we may become immured to. We notice silence when it is framed by sound, when the traffic stops, when Niagara's waters freeze and cease flowing and when the jets from the nearby airport are for some reason grounded. Then, the ensuing silence throws the previously unheard into relief, and we may curse it, or embrace it. To gain control of our listening, on the other hand, and to engage with it as an aesthetic practice, is to enhance and enlarge our participation with the visual world, add a dimension and more enrich all our senses. Active listening 'challenges, augments and expands what we see, without presenting a negative illusion, by producing the reality of lived experience. Through this generative experience listening revisits those philosophical tenets that are bound to the sovereignty of the visual' (Voegelin, 12–13). Above all, I would celebrate the potential of the tiny sound to be remembered, to be added to the mental album like a photograph and to be consciously acknowledged as a powerful force of recreation in a Proustian or Pavlovian way, representing as it does, a doorway or a window into a realm of response and memory.

Notes

1. Picture caption, Walker Art Gallery, Liverpool, UK.
2. Williams, Rowan. 'Reflections on the Vaughan Brothers' in *Scintilla 21— The Journal of the Vaughan Association*, p. 11, 2018.
3. Retrieved from http://www.telegraph.co.uk/culture/art/11731370/What-do-paintings-sound-like.html.
4. Porteous, Katrina, written communication with the author, reproduced with permission.
5. Reproduced by permission.
6. Glanville, Mark in *Times Literary Supplement*, p. 15, 6 April 2018. Glanville was here reviewing *Selected Poems* by the Polish poet Julian Tuwim (1894–1953) translated by Marcel Weyland, pub. Brandl and Schlesinger.

CHAPTER 2

The Notes of Human Music

Abstract We begin with the act of listening, absorbing signals as they reach us, and while they have the voice as their origin, they may not actually be formed of words; a pause or a hesitation may say more of meaning than the language on either side of it. Such traits as dialect and accent become heightened through objectivity. An increased awareness of words as sound can lead us to developed senses of expression, as in poetry. Vocalised sound contains a musicality that has attracted composers, hearing in the abstract a quality beyond, but enhancing meaning. The affectations of speech hide identity while imposing themselves in forms of social power. Meanwhile, the true personality is shown in the physical grain of a voice's making.

Keywords Voice · Language · Poetry · Identity

PATTERNS OF SOUND

External sound comes to us from the past. Light and sound are, as we know, travel at different speeds, so as Michel Chion reminds us, 'hearing – namely the synthesized apprehension of the auditory event, consigned to memory – will *follow* the event very closely; it will not be totally simultaneous with it' (Chion, 13). At the same time, sound is within us. Rhythm, timbre and volume are part of our being, and because we are creatures of call and response—two-way wireless

© The Author(s) 2019
S. Street, *Sound at the Edge of Perception*, Palgrave Studies in Sound,
https://doi.org/10.1007/978-981-13-1613-5_2

communication systems—together they form one of our most intimate means of connection with the world around us. They are fundamental characteristics of life. Our awareness of rhythm, as I.A. Richards has stated, 'is not due to our perceiving pattern in something outside us, but to our becoming patterned ourselves' (Richards, 1924). Rhythm in particular is fundamental to all life, and it literally moves to the drumbeat of our own heart. Sound reaching us from outside sources, including speech, is a noise we decode, and a key part of this process is the location of rhythm that we can interpret intellectually and emotionally. In this, we are linked to the world, just as 'the sea gives birth to a tidal flow...a rhythmic current emerges from the disorderly lapping of waves' (Serres, 120). This is where music begins, heard before meaning emerges, and speech, language is an extension of this, based on rhythm and vibration. 'Whoever speaks is also singing beneath the words spoken, is beating out rhythm beneath the song, is diving into the background noise underneath the rhythm' (ibid.). Jean-Luc Nancy considered that rhythm 'is nothing other than the time of time, the vibration of time itself...' (Nancy, 17), while the great French film director Robert Bresson wrote a note to himself with the reminder that 'nothing is durable but what is caught up in rhythms. Bend content to form and sense to rhythms' (Bresson, 58).

Sound begins as a sub-aural experience, or, we might better say, 'extra-aurally.' We could even pursue the idea to the ultimate and consider that the human expression and interpretation of sound begins with the rhythm of an internal thought. Certainly, sound is visible, as anyone who has used digital editing software will know; lip reading establishes words as visual shapes, just as braille turns words tactile, the difference being that reading speech direct from lips is to engage with real time, the speed and rhythm of speech, rather than its printed form. The percussionist Evelyn Glennie, deaf from the age of twelve, points to the interrelationship between speech, meaning and expression, underlining that communicating is a sensory, bodily act operating on a level far beyond the utility of any one organ:

> When I'm doing the talking, I watch the other person's eyes to see how they are reacting, and when I talk I watch the whole face. If someone suddenly popped on a pair of sunglasses or a mask, I wouldn't be able to follow what they were saying; the eyes are so crucial. Once I stopped

using the hearing aids, I became entirely dependent on what I could read from people's faces and general behaviour when they were talking to me. (Glennie, 45)

Vibration too communicates sound. In Chapter 1, I remarked on becoming aware of activity happening beneath my feet by sound symptoms transmitted through matter, and Glennie herself observed how similar phenomena had been used to demonstrate the patterns of sound at concerts 'where the deaf were offered balloons to hold so they could feel the music through the vibrations' (ibid., 46). It is apposite that we begin a chapter on speech, language and voice with a musical analogy because we build all our empathies and antipathies through observation of all kinds, and it begins with the physical making, recognition and identification of what in its broadest sense is the music of what happens around us. On the very first page of her indispensable guide to the craft of the writer, Ursula K. Le Guin reinforces this idea:

> The basic elements of language are physical: the noise words make, the sounds and silences that make the rhythms marking their relationships... Most children enjoy the sound of language for its own sake. They wallow in repetitions and luscious word-sounds and the crunch and slither of onomatopoeia; they fall in love with musical or impressive words and use them in all the wrong places. Some writers keep this primal interest in and love for the sounds of language. Others "outgrow" their oral/aural sense... That's a dead loss. (Le Guin, 1–2)

As it is for writing and reading, so it is with our reaction to the volume of signals reaching us from outside. Sounds in our mind whisper to us, so a whisper coming from an external source sparks recognition and attracts attention through all the bustle and babble of a crowded room precisely BECAUSE it is a whisper. Our survival antennae become alerted when there is any sense of threat, or equally of self-interest, and that includes potential conspiracy or subversion. What we hear first is perhaps not words as such, but a kind of sibilance, a tone that feels unnatural. Just as students and scholars like 'key' words to stimulate relevance, the mind too is primed to trigger a sense of attention, be it danger or threat, kinship or simply interest, by the sound of certain words that may be generally or specifically significant to ourselves and/or our situation. We neglect the relevance of our powers of subliminal perception at a cost, because:

...the nervous systems of living creatures are subjected to far more stimulation than can be used. On the one hand, the sense organs receive stimuli of great variety. On the other, memories, images and ideas arise internally and must be considered from moment to moment....We are consciously aware of only a limited amount of this information at any moment.' (Moray in Gregory, 75)

Little wonder then that we may be affected by a mood or a feeling that has been created by something occurring, or recently occurred, within our environment for which we have no conscious explanation. A small sound may have a huge range of implications and significances that can affect us without our realising it, and not least of these may be a particular word, or even a sound within that word, or the way in which it is spoken or sung, because 'the awareness threshold for certain words...may be significantly higher or lower than that for more neutral material,' (Dixon, ibid., 885–6) and our personal response is governed not only by our species sense but also by our individual experience and memory. We can consciously manipulate emotion and feeling through such a device; there is, after all, such a thing as the 'stage whisper,' and the so-called cocktail party effect, where we have the capacity to reject unwanted messages delivered through the ears, while focusing on others, a skill depending largely on binaural hearing which allows some signal sources to be localised, becomes increasingly relevant within a cacophonous world. As we filter what we want or need from what we deselect, the quality of the voice, such as the sex or age of the speaker, may also be used to accept or reject signals.

The ears hear, but the mind listens, and as it does so, it absorbs syllables, accidental linguistic patterns, such as assonance, sibilance, rhyming moments between words and sounds heard in passing that attract the attention, even for a split second. All these are the units that make up the music that plays around us and in which we participate, triggering as it does so 'a little storm of felicitous magnetised particles,' as the writer Mark Granier wrote, reflecting on sound in the process of his creative process. It can be 'bits of overheard conversation, in a pub, on Radio 4, on a bus. It's a bat-like sense, to shape a vessel out of sounds/images, the music of what happens.' A poem can form out of such disparate pieces of the sensory jigsaw pieces flying around us, sometimes declaring itself 'out of practically nothing: a chord twangs, an echo bounces, and suddenly there is a little self-contained room, a whole sonic realm' (Granier in Ivory and Szirtes, 15).

Our motor system is activated from the sensory stimuli that our place in the world triggers, and the minutiae of the signals sent and received are sometimes so tiny that they can be overlooked, or not given the credit that their importance—far beyond their apparent insignificance—would at first suggest. The shape and structure of language are made from these sound segments (phonemes) just as the musical score of a vast symphony is created from individual notes of various time values, suggested volumes and emphases. These are 'organized into syllables, with stress patterns or tones; sentences consist of syntactic structures. Each level of the language system is associated with certain types of meanings. Words for example are associated with items, actions and properties; sentences express events and states of affairs,' (Caplan in Gregory, 515) and so on. Yet expression through sound can be more subtle than words. The most complex of arguments, the most sublime poetry and the most powerful oratory are all made up of building blocks, consonants, vowels, phonemes, aspirants, plosive sounds and accidentals that contribute to meaning but, equally importantly, to emotional context.

THE ENERGY OF MICROCOSMS

The link between the visual and the auditory introduced in our first chapter has at its heart, the common denominator of the image. Whether we realise it or not, we desire to make pictures, because pictures are also stories. In the introduction to the book of her *Collected Lyrics*, the writer and performer Patti Smith explains how her creative career began by drawing, but 'standing before large sheets of paper tacked to a wall, frustrated with the image, I'd draw words instead – rhythms that ran off the page onto the plaster. Writing lyrics evolved from the physical act of drawing words. Later, refining this process led to performance' (Smith, 3). Making sense of the World, and ourself, the logical necessity is to make shapes, and the first mark on the page is also the first sound.

It is possible to break down the most complex and sophisticated utterances into component parts, just as a symphony is made of individual notes. These morphemes are the units of linguistic identity, which in turn becomes our individual means of expressing who we are and where we are from. A morpheme cannot be subdivided; it is the first building block. It is not the purpose of this book to explore in detail the complex terrain of linguistics. There are other works that will take this area of study forward for the interested scholar who seeks to learn more of the field. Suffice to say here that we should be aware of a number of factors that identify our similarities and differences in the way we express who we are. So consider what linguistic structure is and how it varies, how language is acquired and operates within the human cognitive system, the function of the brain in production, perception and processing of words, the context for their use in various cultures and how times change vocabulary and pronunciation.

Into this realm comes the examination of pure vocalised sound in the area of phonetics. Articulatory phonetics tells us how speech sounds are produced, acoustic phonetics explores the physical property of sounds such as frequency, duration and volume, while perceptual phonetics (auditory phonetics) is the study of how we actually hear and interpret sound. Phonology, on the other hand, investigates how sounds are organised in a particular language, and in what respects sound systems of different languages relate and diverge.

Yet the human speech when it is uttered rather than written is much more than merely mechanical; learning and hearing language is about nuance and interpretation. The voice can be actual or metaphorical, physical or imagined, personal or communal, political or cultural. Emotion, anger, joy, humour: these all manifest themselves in the human voice and can be interpreted independently of actual language. Tone, timbre, pitch and volume are all factors, but breathing above all drives how the instrument plays the notes given to it by the brain and the heart. Radio presenters are wise to remember that the audience can hear a smile, and its presence may be appropriate or otherwise within the context of what is being said. We can read feeling and mood from the very sound of speech: 'She said she was OK, but I could tell she wasn't. I could hear it in her voice.' It can be even more subtle than hearing a voice break with emotion; a pause or a slight stumble can sometimes be the most eloquent symptom of a subtext that a sensitive listener may understand. Likewise, the pitch or tone of that voice may be untypical of the speaker's normal conversational personality to someone familiar with their everyday expression.

Some years ago, I interviewed an elderly Derbyshire farmer on a visit to the site of his childhood home. The house and garden had long-gone, and in their place, there was just a barren field punctuated by occasional fragments of rubble, but what remained was sufficient for him to recreate the place in which he had grown up. 'This was where the garden wall stood,' he told me as we walked around the muddy field; 'and there was a small orchard about here. And there were lovely apples and pears…that grew.' There was that momentary pause after 'apples and pears,' and in that tiny moment of reflection, the merest hesitation in his speech, it was clear that he was opening a memory that had been buried for more than sixty years. Speech is profoundly animal, and our sophisticated utterances have their origin in modes of survival, courtship, anger and wonder stemming from our earliest beginnings, all borne of air. As the radio producer Eleanor McDowall told me, 'I think more often than not what catches me on playback are pauses in the human voice. Silences, quivering delivery – things that you instinctively feel in the room (but maybe don't hear) which on a second listen can carry a new weight.'[1]

The smallest components of speech as they leave our lips are enough for society to judge us. Listen to the radio: listen to the voice that is speaking. If the voice belongs to someone we do not know, all we have to form an image of them is their voice, what they say and how they say it. Yet electronic media can do strange things to the human voice, once it is divorced from a visual presence. We notice this particularly on the telephone; when a known person speaks to us, there is recognition, but there is often too a separation between the person we know, and can conjure from memory—sometimes intimate memory—and the disembodied voice talking to us. Modern technology has, to an extent, alleviated some of this issue; digital sound is capable of producing near 'picture perfect' audio, so the match between what we know of a person and what we hear converges. Yet it is not long ago that the voice coming over the wires emerged from a sort of electronic nowhere, frequencies and tone flattened, while some other vocal characteristics seemed to be heightened. When I was in the presence of my Irish mother, I was not aware of her Monaghan accent. I was in the presence of her personality, I could see her, and the sound of her voice was only a part of the information I had at my disposal in order to experience the act of our being together. Yet listening to a recording of a telephone conversation now, more than thirty years after her death, there is a strange sense of detachment, almost a lack of recognition. Had I been asked, I would have said that I could 'hear' her voice through imagination in my memory. Yet now, listening to this recording, I need a prompt to remind me that it is indeed her. The main issue is the accent; I hear a very strong Irish accent that I was not aware of previously. There is an objectivity in my listening that was never there when she was alive. I hear the whole landscape of her voice, but reduced and 'thinned out' of some of its frequencies by the recording. What moves to the fore is her dialect, and it comes as something of a shock. This small sonic pattern is made up of tiny undulations and other details, rather like, to use a visual analogy, a landscape seen from a window. Concentrate a sentence, a phrase or a word closely and often enough as pure sound, repeatedly, and it moves away from its meaning and becomes music, and music is made of notes, little microcosms of sound that, placed in juxtaposition with one another, form a shape and propel meaning and feeling through time.

When we meet a person for the first time, we lack the information that familiarity gives us, so all our senses come acutely into play, and leading them is that which we see and that which we hear. Is the voice quiet or loud? That may tell us a lot about personality. How are the words formed? Is there a speech impediment or other vocal idiosyncrasy that will help us identify this person in future? Can we understand them? Is there a strong accent or dialect? Does the voice match the visual image? Often when we meet a person whose voice has previously been our only term of reference, it affects us with surprise. They may or may not look the way we expect them to look, based on how they sounded. Yet when we come face to face with a stranger, we absorb first impressions with the speed of light and the speed of sound, and the whole experience is made up of details. Were we to be listening to this person on the radio, our judgment would be limited but summary. A displeasing voice can be enough for many people to retune or switch off; we cannot 'switch off' a face-to-face encounter in the same way.

It is hardly surprising that the unique instrument in our possession should have been an endless fascination to us; the ability to record human tones caused a sensation when Edison unveiled his invention, but long before this, the idea of reproducing or replicating voice, either through the eerie art of ventriloquism, or by mechanical means, exercised the minds of entertainers and scientists alike. Notable among such devices was The Euphonia, a talking machine invented by the German Joseph Faber and first demonstrated in Vienna in 1840. Faber had been influenced by a book by Wolfgang von Kempelen called *On the Mechanism of Human Speech*, and his device caused a flurry of interest on both sides of the Atlantic, including an enthusiastic response by the American physicist, Joseph Henry, who observed that its operation, using sixteen levers or keys 'like those of a piano,' enabled the equivalent number of elementary sounds by which 'every word in all European languages can be distinctly produced.' A further key opened and closed the equivalent of the glottis. 'The plan of the machine is the same as that of the human organs of speech, the several parts being worked by strings and levers instead of tendons and muscles.'[2]

In 2018, Faber's creation provided the title for a research-based project by the artist, Emma Smith, developed across a number of UK galleries, beginning with the Bluecoat in Liverpool. Smith's preoccupation, however, differed radically from Faber's artificially created vocalisings; her premise that conversational interaction creates harmonies and interlocking rhythms working musically on a subconscious level drove a series of experiments using real voices from a range of languages—English, French, Italian, Japanese and Hindi—to create a score that could be then played as part of an installation. Sounds coming through the vocal chords create pitch, which is a note in the musical sense: thus, human speech is music. Choral singing in its traditional community form is known to provide a sense of belonging and connectivity, but Smith arrived at the idea that there is music in speech and voice not reliant on learnt material but on improvisatory interaction through a process that is subconsciously collaborative: we match pitch and rhythm in conversation without realising it as we talk, making music with one another intuitively through conversation.

> The musicality of the actual syllables and the sounds that we're using, they also play a part. There's also this complex poetry that's going on in conversation where, when I'm wanting to choose the meaning I give to you, I'm making decisions: what are the words that give you the meaning, how can I put these words together to give you the music in the way I pronounce them, fluctuate the voice or add rhythmicity? And it is all happening subconsciously at super- high speed level that we're not really aware of. Meaning and music are heavily entangled in ways that are more communicative than we think.[3]

Later in the chapter, we shall return to the creative aspect of speech as pure music. There is no doubt, however, that the preoccupation with capturing the voice for reproduction has played a vital part in our ability to analyse in this way. At the turn of the twenty-first century, the BBC ran a major project which sought to take a 'snapshot' of the everyday speech and speech-attitudes of the broadcasting audience within the UK at the time. It was called appropriately *Voices*. As editors Clive

Upton and Bethan L. Davies explained in a book which accompanied the project, 'it was seen as informing popular understanding of English and other languages, generating programmes for national and local radio and television, and material for the BBC website and publications' (Upton and Davies, xii). Radio creates a heightened consciousness of speech and voice, and broadcasting—particularly British broadcasting—has been obsessed with the voice from its earliest beginnings. To hear a voice, in particular our own voice, is to become extremely self-aware, and self-consciousness of sound's effect on others is a fundamental part of society, in particular in areas where language is deliberately heightened, as in drama and poetry. Often actors are criticised for *performing* the reading of a poem. At the other end of the scale, the poet themselves may internalise their own work, to the detriment of the audience. A good speech teacher or voice coach has the capacity to isolate issues and break down problems of communication into component units, the microcosms of meaning.

WORD PICTURES

The poet Peter Levi wrote that 'poetry is language heightened by insistent sounds or repeated rhythms' (Levi, 30). This can be part of the conscious creation of a work on the page, but equally it can be an unconscious response to a powerful set of emotions or circumstances. It is not the exclusive domain of the professional poet or writer; listen to an eyewitness to a disaster or someone impassioned in a debate on a subject about which they feel very strongly. At a certain point, their language may become transcendent, and, questing for adequate means of expression, they use words and sounds in a way that itself becomes poetic. Orators and writers simply develop such innate abilities deliberately or their purposes. 'Once you use language consciously and intently you cannot escape the special demands that are inbuilt into poetry' (ibid.). The manipulation of sound is often demonstrated by such poetic devices as alliteration, assonance and onomatopoeia, the formation of a word from a sound associated with what is named, such as 'cuckoo' or 'sizzle,' and famously in William Wordsworth's description of skating in his long autobiographical poem of his youth, 'The Prelude':

All shod with steel,
We hissed along the polish'd ice... (Wordsworth, 27–9)

The English Romantic poet John Keats wrote a perfect composite of pictorial description in stanza XXIV of 'The Eve of St. Agnes' in which individual sounds and words made up of image-creating vowels and consonants contribute cumulatively to create word colour of extraordinary detail and richness. When the poet's contemporary, the artist Daniel Maclise (1806–1870), painted 'Madeline After Prayer' (Walker Art Gallery, Liverpool), he depicted the scene as described in Keats's poem, but even the precision of his interpretation cannot match Keats's own word picture, made up as it is of intricate minute internal sounds that transmit from the page into the reader's mind as images:

> A casement high and triple-arch'd there was,
> All garlanded with carven imag'ries
> Of fruits and flowers, and bunches of knot-grass,
> And diamonded with panes of quaint device,
> Innumberable of stains and splendid dyes,
> As are the tiger-moth's deep-damask'd wings;
> And in the midst, 'mong thousand heraldries,
> And twilight saints, and dim emblazonings,
> A shielded scutcheon blushed with blood of queens and kings.
> (Keats, 201)

In his poem, 'The Song of Hiawatha', the writer Henry Wadsworth Longfellow created a vast poem of Native American life and myth, using throughout its 198 pages repetition, sound colour and above all an unfaltering rhythm simulating the sound of tom-toms that demand to be given physical voice:

> Whence these legends and traditions,
> With the odors of the forest
> With the dew and damp of meadows,
> With the curling smoke ofwigwams,
> With the rushing of great rivers,
> With their frequent repetitions,
> And their wild reverberations
> As of thunder in the mountains? (Longfellow, 113)

Little wonder this huge poem has lent itself to performance and dramatic interpretation. Spoken texts have the quality of emotional immediacy, but they must communicate meaning at first hearing: the page cannot be turned back for a second reading in a darkened theatre. Sometimes, however, the syllables have no literal meaning, yet convey sense, colour and image through pure sound. The English poet John Clare, confronted by what he sometimes felt was an absence in language for his poetic purposes, was not above creating new words, the sound of which he felt fitted more exactly the sonic/visual image he was seeking to convey than that which the more orthodox vocabulary available had to offer. Likewise, the nonsense verse of Lewis Carroll could be eloquent in the extreme and calls—as in 'Jabberwocky'—for the voice of a pseudo-melodramatic orator:

> 'Twas brillig, and the slithy toves
> Did gyre and gimble in the wabe:
> All mimsy were the borogroves,
> And the mome raths outgrabe. (Ricks, 190)

Charles Darwin observed wryly that 'the impassioned orator, bard, or musician...little suspects that he uses the same means by which at an extremely remote period, his half-human ancestors aroused each other's ardent passions' (Darwin, 330).

WORD MUSIC

Far from such conscious contrivance, in everyday language, even the imperfections of colloquial speech are eloquent. Talking spontaneously, we stumble, stammer, seek to find the right word and pause as our brain explains or digests the meanings behind what we are trying to express. As Craig Dworkin has written, 'statistically between 7 and 10 percent of all speech is dysfluent, with phonemes repeated, prolonged, distorted, suspended – or even at times, not audibly produced at all. The idealogy of transparent and referentially communicative language is so strong, however, that we tend to overlook those dysfluencies or not consciously register them in the first place' (Dworkin in Perloff and Dworkin, 166). Sometimes controversy may be provoked by the radio producer's skill in editing our speech for broadcast, taking out these imperfections in the interests of clarity and often indeed temporal expediency. On the one hand, an argument may be that he or she is freeing the language of defects or obscurities of meaning and therefore clarifying the speaker's intention, while on the other the view may be taken that in the editorial process language becomes manipulated out of the speaker's control. Having said that, if we discount the relatively modern phenomenon of recording and playback of sound, or even the mass-produced ability to copy the marks that are representations of sound in the form of words and music on the page, we return to the basic facts of pure sound as communication in a form where the page cannot be turned back, the tape cannot be rewound, and the honed pen of the senses carves the immediacy of passing experience into memory. Now, those individual exclamations: 'Ah!' the scream of pain, the sob of sorrow, the laugh—so full of many meanings from joy to derision—all the human sounds that communicate without translation, these become vitally important. We have learned to 'hear' them, even when confronted with their notation; we inherit the sounds and languages together with their rules and forms from previous generations. As I grow older, I can hear my father's voice in mine, not in the overall sonic impression of it, but in small idiosyncrasies of tone, pitch and rhythm. A great playwright such as Shakespeare captures not only profound expressions of meaning, but also the subtleties of sound that make us human and individual. 'His enormous vocabulary and his very strong sense of the sound and behaviour of English make him a kind of embodiment of the language itself' (Levi, 30).

It is as Claudia Daventry has written of the ambiguity within living language that 'not only its images but its very words have the inner heat of their individual histories – the fossils within which, like coal or oil, release the energy of their original iconic power when ignited' (Daventry in Cambridge, 174).

The American composer Steve Reich has been interested in speech melody since the early 1960s. It is particularly pronounced in the voices of children, but there is a kind of music that can be created by mixing animated vocal sound with more monotonous speech. Language, that is to say the native language of the speaker, be it English, Chinese, Arabic, Hindustani or any other, affects all vocal music created in that language, through its particular cadence and rhythm, the model for this being folk music, which 'tends to be a very direct setting of the language in its common vernacular form' (Reich, 194).

....The most important musical aspect of language [is] its sound. As a composer, if I don't care for the *speech melody* of someone speaking, then the meaning of their words is of no consequence. On the other hand, once the speech melody has caught my ear, the meaning of the words can never be overlooked. (ibid., 199)

We shall return to the issue of speech as music in a later chapter. Nevertheless, it is worth stating at this point that the relationship between words has demonstrable connections, insofar as some of the neural processing involved has significant points in common in terms of analysis. 'Understanding speech, for example, requires that we segment a flurry of sounds into words, sentences and phrases, and that we be able to understand aspects beyond the words, such as sarcasm (isn't *that* interesting) through stress and other emphases such as volume' (Levitin, 86).

The construction of dramatised language to convey empathy in a listener and/or audience combines heightened word and phrase sounds, poetic rhythms and repetition and recognisable tropes of everyday speech, in a blend that can convey complex ideas within a framework of internal shapes and signifiers that create an overall extended performance piece. There is clearly a link with the poetic here. In the work of the poet and playwright, T.S. Eliot, the use of sound, and in particular his theories of 'auditory imagination' as exemplified in his long poem, *The Waste Land*, and its links with the contemporary world in which it was created, will be instructive to consider as we seek to break down the song of popular culture and language usage into some of their key components.

The Wind Under the Door

Sound—like life itself—exists in time; we catch it as it passes us, and as we do so, it is already fading, moving on and away from us, becoming memory. Like dance or abstract painting, pure sound can act on our senses to open doors to individual worlds of imagination. Sound is the fundamental medium of storytelling. Its transmission and reception are innate skills that we all possess, and yet this kind of intense poetic response-experience requires attention, the part-learned, part-intuitive discipline of active listening and responding, tuned by instinct as well as by the search for meaning. Words—and the spaces between them—can transmit pictures physically through the ears to the brain in concert with the eyes, but we are capable of consuming signals from a silent object, such as a text, through the imagination. *The Waste Land* is rather like a box of disparate cultural fragments in which Eliot gathers the images of a broken society and sews them together through connections between historical and national identity. As readers, we absorb this babble of voices and sounds as receivers, and it is this 'bed' upon which perceived images settle. He wrote of the sound of words as if he were moving a tuning needle across crowded radio bands. There is music (a snatch of Wagner, a thrush's song and a nursery rhyme among others) yet almost always there is a voice at the heart of things, even when what we are hearing is elemental, because the listening is to Eliot's sound pictures, through the apparent silence of black words on white paper. The imaginative digestive system in which the music of Eliot's meaning is interpreted through sound which—because it travels through time—evokes moments that once written, remains fixed in rhythm and syntax.

In February 1942, Eliot delivered the W.P. Ker Memorial Lecture at Glasgow University. He chose as his subject, 'The Music of Poetry.' His talk was wide-ranging, but key to the argument was the relationship between poetry and conversation. Listening to a speaker in a foreign language, the music of conversational tone can convey a sense of meaning, provided there was meaning behind the intention as opposed to plain gibberish. Likewise, to hear a voice speaking in dialect is to hear the music of identity, even before meaning is interpreted. A poet, said Eliot, 'must, like a sculptor, be faithful to the material in which he works; it is out of sounds that he has heard that he must make his melody and harmony' (Eliot, 17). Since the emergence of radio, the spontaneity of the spoken word had become ubiquitous, and a consciousness of the grain of the human voice—timbre, dialect, euphony and rhythm—and the geographical and cultural variety of these elements informed a generation of writers' work. Eliot saw in poetry a point where language could move from speech to song, linked by the one basic element of human communication, that is to say, voice. Words after all were sounds before they were marks on paper, papyrus or stone. Eliot coined his phrase, 'the auditory imagination' in a book of essays published in 1933, entitled *The Use of Poetry and the Use of Criticism*,[4] to which Helen Gardner in her key text, *The Art of T.S. Eliot*, directs us, in particular to a key passage where Eliot defines his term specifically as 'the feeling for syllable and rhythm, penetrating far below the conscious levels of thought and feeling, invigorating every word; sinking to the most primitive and forgotten, returning to the origin and bringing something back, seeking a beginning and an end' (Eliot, quoted Gardner, 6). Language in whatever form, and however sophisticated or not, through word or explication, phrase, syllable or complex compound sentence expressing thought, 'remains, all the same, one person talking to another; and this is just as true if you sing it, for singing is another way of talking' (Eliot, 16). In listening to the world, we must attend to the sounds that are closest to us, the very sounds of ourselves. 'Thus,' writes Arthur Koestler, 'rhythm and assonance, pun and rhyme are not artificially created ornaments of speech; the whole evidence indicates that their origins go back to primitive – and infantile – forms of thought and utterance, in which sound and meaning are magically interwoven, and association based on other similarities' (Koestler, 315). The hissing of the wind under the door is nothing less than the origins of our sonic beginnings, insinuating themselves into our awareness as we turn our attention consciously back upon them: 'Rationality demands that these matrices should be relegated underground, but they make their presence felt in sleep and sleeplike states, in mental illness and in the temporary regression – the *reculer-pour mieux-sauter* – of poetic inspiration' (ibid.).

Language, Caged and Flying

We are what we hear, so the beginning of language is in listening. Clearly, this is the case with the growing child, but we should remind ourselves that this process never stops. 'To speak in my own voice, I have to fall silent and listen...In this brief absence, I begin to find my own rhythms' (MacKendrick, 35). There is an analogy here with writing; a developing poet or novelist who is serious about their craft understands that reading is the doorway to writing, but too seldom it is practised in a way that informs individuality. To bring these two allied acts of reflection together would be to re-engage with a past custom that is of reading aloud. As MacKendrick says:

> Voices come to us in the way our bodies echo and alter: we read, we hear; we acquire an accent...The embodiment of voice demands that we attend to the sonorous sense even of a text given visually. Sometimes it makes sense to read aloud; other times, at least to subvocalise...Not only should we be surrounded by words as by touch...but someone, from the first, should enwrap us in the way that words on the page work in and on the voice. (ibid., 37)

The parent or guardian of a growing child is in a unique position to do this from birth, but the idea of sharing words aurally in school or college is a neglected practice that has real relevance to an awareness of sound as vital human expression. To roll a word around in the mind and then feel its effect on the mouth as it is spoken, plosives, their impact on sense and feeling, the interaction between voice and wall, the echo and reverberation of human-made sound across acoustic spaces, is immediately to be consciously present in Place, and to demonstrate that presence, while learning to understand that this presence is a crucial part of identity. Even in darkness, it is a rather beautiful mystery that voice is sound made by a body, physically created out of body parts, and yet it is invisible (unless interpreted graphically). 'The voice is the element which ties the subject and the Other together, without belonging to either, just as it formed the tie between body and language without being part of them' (Dolar, 103). It is sometimes a revelation for someone to actually see vocal patterns on a screen, say on a digital sound editor, because the impact of every minute aspect of human sound suddenly becomes visible and as near as possible, tangible. The pauses, the stammers and stumbles, emphases and murmurs that make up a single word, they are all there to see; we are suddenly aware of cause and effect in a way we never were before. We have spoken of the power of a speech radio producer to manipulate in a way that may be almost imperceptible and yet which can produce a telling effect. Likewise, we have the technology to change a pitch or the tone of a singing voice, or even to shorten durations without affecting either of these. In this field, nuance is everything.

There may also be another problem here, and this belongs to the listener. When we hear a voice for the first time, it may be that the first words are lost to us; it is perhaps analogous with being introduced to someone for the first time. We hear a name, but we do not always retain it; our senses are working so hard at processing information about this new presence that their actual name may pass us by. 'What a lovely person!' we may say afterwards. 'What was her name again?' To overcome this, it is a conscious strategy to repeat the name as it is first heard, in other words, to take ownership of the sound through our own apparatus and thus help commit it to memory and association. A voice and/ or a dialect may be highly distinctive, but it can be too much so at times, and sound can overmaster meaning. Or it can convey it without the aid of language. Robert Frost, the great American poet, called it 'the sound of sense'. Of this concept, Frost wrote: 'The sentence sound often says more than the words. It may even, as in irony, convey a meaning opposite to the words' (Frost, 113). Likewise: 'A sentence is a sound in itself on which other sounds called words may be strung. You may string words together without a sentence-sound…just as you may tie clothes together by the sleeves and stretch them without a clothes line between two trees, but – it is bad for the clothes' (ibid., 110–11). Frost's sound of sense is linked to 'that particular kind of imagination that I cultivate rather than the kind that merely sees, the hearing imagination rather than the seeing imagination...' (ibid., 130).

Thus, the sound of speech as affected by dialect, environment, state of health, living habits, speech imperfections and all the other symptoms of living governs our expression. Edit us too harshly, and our friends and family would not recognise us; transcribe or translate our words onto a page carelessly, and you risk turning our thoughts anonymous. Some years ago, I made a radio documentary about the work of the Sound Conservation Department at the British Library in London. While making the programme, I heard for the first time, voices of working men from before the First World War, soldiers voices, the voices that my grandfather and grandmother—and their parents—would have heard. The extraordinary thing about these recordings is that they are voices in captivity; it is an audio archive made between 1915 and 1938 by German sound pioneer, Wilhelm Doegen. In this remarkable project, Doegen sought to capture the voices of people—their languages, music and songs—from all over the world. The collection acquired by the British Library in 2008 comprises 821 digital copies of shellac discs held at the Berliner Lautarchiv at the Humboldt Universität. It includes recordings of British prisoners of war and colonial troops held in captivity on German soil between 1915 and 1918 and later recordings made by Doegen in Berlin and on field trips to Ireland and elsewhere. The content of the recordings varies and includes reading passages, word lists, speeches and recitals of songs and folk tales in a variety of languages and dialects.

As I listened, I heard the voice of a Durham man reading from the Bible in 1915. I shall never know who he was, but he exists for me now, because I heard his voice. We shall return to this poignant concept of human identity in a later chapter. The German recordist used the same passage—the Parable of the Prodigal Son—as a control, so it was possible to analyse speech differences within the same sentence structures. Not only English speakers were recorded; here are British Colonial troops, speaking in their mother tongues, and Welsh soldiers speaking Welsh. I listened to the voices with Jonathan Robinson, curator of social science at the British Library. It becomes clear as one listens that the German team was focusing on particular strata of vocal expression:

It's interesting that there seems to have been no attempt to capture what you might call officer class voices; it was clearly the regional accents that he wanted. Among the most interesting of these audio ghosts is the voice of a soldier from Bletchington. This rural village is in Oxfordshire, about 60 miles from the capital. It's so close to London that today it's barely perceived as having an accent, but I think people would be startled to realize how "West Country" the accent of rural Oxfordshire sounded at that time.[5]

It is sound as witness from the early days of recording, and while it is fascinating, it is also tantalising, because while listening, one cannot but think of the millions of lost voices that came before. We know language evolves, and while we may recognise the individual notes and understand their meaning, to hear evolution in a voice is to hear the infinite variety of human music through history. How we speak is informed by where we are from. In the UK alone, there remains a great range of spoken styles, and even by travelling around such a relatively small set of islands, we may become aware not only of this variety, but also of the prejudices of those who speak in a particular way. The 'correct' way to speak British or American English is only one example of these entrenched views; everyone has their own opinion as to what is right. Let us consider just one word, perhaps an appropriate one: 'class.' Historically, originally, the word would have been pronounced with a hard 'a' sound, but a regional change began about 400 years ago, and speakers in certain parts of the UK, in particular in the London area, began to say the word with a soft 'a' as in the phonetic pronunciation, *klahs*. The hard 'a' persists in the north of the country today, because of its geographical distance from the capital, and thus, it is more resistant to change, even now, when influence comes not only through ease of travel, but also through media.

It is interesting to note that the idea of 'Received Pronunciation' ('R.P.') sometimes referred to as 'BBC English' because of the style of speech adopted by early national broadcasters—clipped and affected in its worst examples—was only spoken by approximately 2% of the British population. When it was first heard over the airwaves during the 1920s and 1930s, it became identified as that peculiarly British phenomenon, the socially prestigious voice, the voice of authority and power. Listening to recordings from some public service announcements today can be amusing: *band* becomes *bend*, *mad* becomes *med*, *hat* becomes *het* and so on. Even in its less-ludicrous extremes, remnants of this persist in certain strata of society, not a true version of the spoken language but a manufactured one. 'R.P.', although a minority voice remained a symbol of power and influence for many years; sometimes referred to as 'King's' or 'Queen's' English, it became replaced by the term 'Standard English' in the 1960s and 1970s with a sound that was not so affected, although the term 'R.P. seems to have returned in some quarters as an expression, if not any longer as a true description. Be that as it may, School and university, family background, social connections and formative environments govern how we sound, and as a result of this, how we sound may ring-fence us within our 'class.' This in turn may decide career paths and opportunities in life.

WITHIN THE GRAIN

When Roland Barthes listened to a Russian bass singer, a man in a church intoning a sacred text, he heard something beyond—or even before—the meaning of the words from the text the performer was interpreting and communicating, something even beyond style and form; it was something that *was* the cantor's body, 'brought to your ears in one and the same movement from deep down in the cavities, the muscles, the membranes, the cartilages, and from deep down in the Slavonic language, as though a single skin lined with the inner flesh of the performer and the music he sings' (Barthes, 181–2).

It is this that we need to carry forward into the next two chapters of this book: the *voice* of *things* happening in their place, the voice that 'bears along *directly* the symbolic over the intelligible, the expressive. The "grain" is that: the materiality of the body speaking its mother tongue' (ibid., 182). By 'mother tongue' within the context of this discussion, we are actually moving beyond even language itself, into pure sound as it reaches us, and as we find ourselves transmitting itself to our self and to the world, in the form of everything down to the smallest sigh. It is sound before it is processed, but it is essentially living. Barthes, listening to music in recorded form on LP records, was critical of the effect of technology that while preserving a moment, somehow sterilised it. How much more would he bemoan the less-desirable qualities of compressed digital sound through which our music is so often consumed today? We are provided with information, but emotional and cultural communication is something other than this, so we are required to actively engage in imaginative reconstruction to recover the 'grain' from sound carriers and platforms in which 'the various manners of playing are all flattened out *into perfection*: nothing is left but pheno-text' (ibid., 189).

Barthes' 'grain' lies at the heart of this study; it is 'the voice as it sings, the hand as it writes, the limb as it performs' (ibid., 188). In other words, it is an essence of being, and it is elusive, but recoverable within a single note of a song, an emotion transferred to the page, and the brush and rush of invisible tides on a beach at night.

To reach that kind of essence, we must take music to mean a sense of almost all heard sound that reaches us, including its vibration beyond the purely auditory, 'to displace the fringe of contact between music and language' (ibid., 181), the 'very precise space (genre) of *the encounter between a language and a voice*' (ibid.). Formal music has its notation, and spoken speech has its phonetics. In language, if a monophthong is a vowel that has a single perceived auditory quality, then a diphthong—a sound formed by the combination of two vowels in a single syllable, in which the sound begins as one vowel and moves towards another— introduces, in the most subtle way, melody into speech. These are minute points of focus and concentration, but they lie at the heart of things in our microscopic sonic world, and they move us between species of music and so into our next chapter, and our exploration of fragments of sound that are in themselves complete.

NOTES

1. McDowall, Eleanor. Correspondence.
2. Computers and Computing—Automata: Joseph Faber. http://history-computer.com/Dreamers/Faber.html.
3. Smith, Emma. Interview with the author.
4. Eliot, T.S. *The Use of Poetry and the Use of Criticism*. London: Faber and Faber, 1933.
5. Street, Seán. Walls of Sound. *BBC Radio 4*, 2011. http://www.bbc.co.uk/programmes/b00zq9mz.

CHAPTER 3

Making the Moment Singable

Abstract The composer Ernest Bloch leads us into a dark wood, and we explore the relationship between the physical creation of music and the act of listening. We turn our attention from the notes of speech to the reverberations of musical notes in space; here can be transcendent moments that seem timeless within small phrases of music. The music of the English composer Jonathan Harvey makes its sounds from bells and from birdsong, as did the French master, Olivier Messiaen, who transcribed sounds from the natural world for piano, and held Time in music made in the darkest of times. The capacity of music to evoke distant times and places: Janet Cardiff's sound artwork, *40 Part Motet*. Human birdsong may be folk song, created out of the oral tradition.

Keywords Music · Birdsong · Composition · Sound art · Folksong

MATERIALISATION OF THE IMMATERIAL

As we remarked at the outset, visiting an art gallery or museum is an auditory experience even before specific artworks are examined; every room has its own acoustic, and the cumulative memory of such a visit is often retained as a sonic echo in the mind. The environment provides a series of circumstances through which the visitor moves, hearing the acoustic of place, ever changing as the crowds move and shift and then focusing attention on specific art pieces. The Russian composer Modest

Mussorgsky, in his work, *Pictures from an Exhibition*, provides us with a clear musical interpretation of this experience, juxtaposing as he does a theme representing the *promenade* of the visitor between individual artworks themselves, each of which Mussorgsky represents in sound. Mussorgsky's work was created after visiting a memorial exhibition of work by his friend, the young Russian painter Viktor Hartmann, held at the Imperial Academy of Arts in Saint Petersburg during February and March 1874. The exhibition contained over 400 of the artist's works, and Mussorgsky's suite, made up of ten sonic representations of individual paintings, was completed in just over three weeks, from 2 to 22 June 1874. In this chapter, we will explore the capacity of circumstance to initiate moments of musical expression. We may imagine Mussorgsky pausing in front of a picture, gaining musical inspiration from suggestions triggered in his mind by the image. Here, we will seek to reverse the process, by considering sounds that are themselves pictures, notes that may be as small as a raindrop and yet which reverberate within a space, be it a physical place such as a church or a garden, or within the mind as imagination and/or memory. The writer, artist and performer Patti Smith wrote that 'we all have a song. A song comes spontaneously, expressing joy, loneliness, to dispel fear or exhibit a small triumph. We hardly notice we are forming them, as we sing them, often alone, half to ourselves. It is finding the words within that leads us to sing' (Smith, 1).

One of the great mysteries and wonders of music is that its creation requires such a combination of the psychic and the physical; there is listening of course, but there is also the physical act of the musician as maker in bringing the sound into the world, listening themselves as they do so. Barthes spoke of 'a muscular music in which the part taken by the sense of hearing is one only of ratification, as though the body were hearing' (Barthes, 151). At the same time, there is a kind of pale light of silence out of which the first sounds a composer hears will emerge, because they will always be the first person to experience that fragile moment. What is then transmitted to us as an audience is a recreation of that moment filtered through performance, and yet, when it touches us most profoundly, something beyond that occurs, perhaps taking us in someway back towards the strange pale silence of the origin of the sound itself. As Barthes says, 'the truth is perhaps that Beethoven's music has in it something *inaudible*' (ibid., 152).

The Swiss-American composer Ernest Bloch once had a dream:

We hear only ourselves...Whatever we shape leads back around ourselves again...We walk in the forest and feel we are or might be what the forest is dreaming... We do not possess that which is all around us...because we are it itself and are standing too close to it, the spectral and still ineffable nature of consciousness or interiorisation. But the sound burns out of us... (Bloch, 1)

These words are how Bloch prefaces his essay, *The Philosophy of Music*, in part based on his 1918 book, *Geist der Utopie*. He was born in Geneva in 1880 and died in Portland, Oregon, in 1959, at the age of 78. Two years before his book was published, he had moved to the USA, where he had taken a teaching post at the Mannes School of Music on Manhattan's Upper West Side. His writings, collected and translated in 1985 by Peter Palmer under the title, *Essays on the Philosophy of Music*, show us a man for whom spirituality and even esotericism were linked to practicality; he must have been an inspiring teacher. After the prelude of his 'dream,' his essay continues with an equally arresting opening sentence: 'How do we hear ourselves first and foremost? As endless singing-to-oneself...' (ibid.). Note by note, as word by word, meaning unfolds itself. In this chapter, we shall explore the constant immediacy of sound in the present moment and its place in the centre of the heard world. All sound happens in time and air, space and place, as we do, and the dramatisation of that lies in music, which, as David Byrne has written, 'far more than being merely entertainment...is a part of what makes us human. If the beginning was with God's Word, as the Christian bible states, then this signifies a sonic moment, an instant; whether it is metaphor or statement of reality, it implies sound, be it the Big Bang or a subtle but all-pervading 'celestial vibration' (Byrne, 323). I have written elsewhere about music and memory, and explored the constant present tense of music, and how this can help us in the most intense predicaments of mental bewilderment, when memory fails us and we do not know who we are. Then, the note-by-note *presence*—always in the *present*—of music touches us as it passes, and we are in the world again. Bloch's dream teaches us that, because in the midst of the dark forest we hear ourselves and in the split second of a crotchet or a quaver, we become at one with the sound.

We overlook the tiny things at a cost; the giant painting dominating the room draws the eye, and we pass by a perfect tiny miniature on our way to it, yet all is only a matter of scale. Sometimes the senses are required to expand in order to understand the minute. Seated at the piano, a writer of music may pick out a series of notes gathered from the imagination. It may be a simple phrase: something that might be the germ of a giant symphony. On the other hand, it may be the work itself, just as a haiku of a few syllables can contain everything that needs to be said. Such is the essence of the tiny *Bagatellen* of the Ukrainian Valentin Silvestrov. During 2006, during a series of recording sessions at the Himmelfahrtskirche (Church of the Ascension) in Munich, Silvestrov would enter the building, very early in the morning, before the engineers and other staff arrived, and sit at the piano. Gently touching the keys, he noticed how the emerging small notes grew great in the giant acoustic, taking shape in the moment. For Silvestrov, these were not fragments or preludes but whole and complete works of great profundity and meaning. The sound of music at its absolute birth has no owner, belonging as it does to the oral tradition of the universe: the language of music lies in the moment of the word of its utterance, irrespective of who speaks or sings it, regardless of where or when it is spoken or played. Indeed, Silvestrov, alone in the great church at the keyboard, might have added that the sounds at his fingertips belonged—if they belonged at all—as much to the voice of the building through which they reverberated. The orality of unwritten music is part of its essence, a key part, although its significance is often overlooked. The folk song passes from mouth to mouth, and as it does so, it is owned and changed by the personality, memory and mood of its current curator. The Austrian nineteenth-century Romantic composer Anton Bruckner improvised at the organ at St Florian, and these improvisations turned into his symphonies. Louis Armstrong's transcendent trumpet solo towards the end of 'West End Blues', although lasting only around 34 seconds, makes a moment in Chicago on the 29 June 1929 stand still. Jazz may be music that changes every time it is played, but in a sense the same is true of all music, of all live performance. Committing a work to paper does not rob it of its potential for growth and development through interpretation, and this interpretation is not limited to the role of the performer, because the ability to actively listen creates the potential for a new experience every time a work is attended to. Some minute instant in a song heard a hundred times can, on the one hundred and first hearing, reduce us to tears. Music is past, present and future heard at one and the same moment. The musicologist Horst Weber wrote that it was a slightly cautious publisher who invented the term *Moments Musicaux* for Schubert's op.74

piano pieces of 1828, because there was no other criteria for them, yet in doing so, 'he lit on part of what happens in Schubert's music – namely, a moment is given duration. Schubert thereby developed a new form of musical time, which emerges not so much from the linking of notes into thematic discourses but rather, in a manner of speaking, from the sounds of the notes themselves' (Horst Webber in Frumkis, 18).[1]

The song without words is a song that can be 'sung' on any instrument or by any voice, conveying a unique and changing meaning everytime, while retaining faithfully its original concept, expression and identity. Once captured, the moment becomes singable ever after. In Valentin, Silvestrov's *Bagatellen* exist a remarkable synthesis of improvisation and composition: 13 tiny works lasting in some cases little more than a minute, and with a total playing time of about 34 minutes, in which the composed element is not always immediately audible at all. These perfect sound shapes, examples of what Silvestrov calls the '"materialisation of the immaterial"...cluster together in colonies or families, in which each is borne of the others and all are related...they form a sort of improvised cycle played in a single breath, a cycle whose basically ceaseless flow is framed by the repeat of the second piece.' Yet what we hear is not improvisation in the strict sense and only seemingly 'oral music: everything has been fully crafted in the composer's mind down to the nethermost detail' (Frumkis, 19). Silvestrov's text is on the surface, quite simple: little development, no dazzling technical devices or timbral effects, but a paucity of notes, full of space and the softest of pianissimos on the very edge of the audible that seem to possess the capacity for infinite expansion within the interior of the church that provides their sounding board. It is weightless music that affects the ear like the impressionistic shadows on a daguerrotype. It the essence of what this entire book is based upon, just as are the oft-quoted lines by William Blake in 'Auguries of Innocence':

> To see a World in a grain of sand
> And a Heaven in a wild flower,
> Hold Infinity in the palm of your hand
> And Eternity in an hour. (Blake, 67)

It is music 'as a collection of resonances, as a response, a postlude to things already said, sung or sounded' (Frumkis, 20). All sound may pass through time, but there are moments that seem to become timeless and turn to something other than the sonic, to light, colour, to the otherwise inexpressible. There are flashes of light in Silvestrov's music that touch something momentarily, like a yearning for an unreachable past, but they are so fleeting that this quality is all but ungraspable. There is something, for example, in a sustained A-Major note played softly on strings—as in Silvestrov's Sixth Symphony's third movement, as in Mahler's opening to his First Symphony, that is akin to a memory of gentle wind in the tops of pine trees, heard through a glimpse of a lost world of childhood. This is impressionistic sound painting of the simplest most elemental form, and because of its lack of embellishment, it contains the power to bypass analysis. Sound's capacity to provide Proustian lightning instants of recollection is the key to a mysterious door. We might say that the power of music itself is in its essence unsolvable, and we may then add that 'is an answer always really required when you can have a beautiful question?' At the same time, we might be curious enough to probe a little deeper into the dense wood.

What the Forest Is Dreaming

Ernst's Bloch's poetic evocation of the source of our relationship with music, quoted at the start of this chapter, is part of that key, but as his dream implies, the context for flashes of sudden light is a sombre dusk. Composers such as Silvestrov are tapping into something that seems to come through them rather than from them, and the strangeness of the sounds communicated from an almost silence is as palpable for the maker as for the listener, because, in a sense, both maker and listener are the recipients of signals from far shores, passing from one to the other like split-second transmissions.

How we actually process all this in neural terms is both complex scientifically and mysterious culturally. It is not easy to trace, because the brain deals with different aspects of sound in different areas, with involvement in musical activity occupying virtually every subsystem. Thus, 'the brain uses functional segregation for music processing, and employs a system of feature detectors whose job it is to analyze specific aspects of the musical signal, such as pitch, tempo, timbre, and so on' (Levitin, 86). The mystery lies in not the brain, but the mind and its emotional servant, the cortex. What Levitin calls 'regional specificity' within the brain is crucial to our mental and cultural absorption of sound. As he says, 'the brain is a massively parallel device. There is no single language centre, nor is there a single music centre' (ibid., 87). However, relatively recent research demonstrating the potential for reorganisation of these specific areas in terms of mental crisis (a phenomenon encapsulated in the term, neuro-plasticity) shows the specificity may be temporary 'as the processing cen-tres for important mental functions actually move to other regions after trauma or brain damage' (ibid.), evidence that explains to a point some of the mental adjustments made during the extraordinary example of the musician and broadcaster Clive Wearing.

I have discussed Wearing's remarkable case elsewhere,[2] but it is appro-priate to briefly revisit it at this point. In March 1985, Clive Wearing became afflicted with a condition known as *herpes encephalitis*, a virus that caused inflammation and subsequent damage to his brain and in particular the areas associated with memory. Wearing lived in an almost continuous present, retaining only a few seconds of recollection at a time, surrounded by a fog of dense amnesia. One of the reasons that his case became celebrated was the role music played in his life, both before and after the disaster of the virus, because, as a musicologist and choir leader, his responses to sound could be observed in a practical and demonstrable way. On one occasion, at a reunion of his choir, this was shown in the most moving way, as recorded in his wife Deborah's book, *Forever Today*. The minute individual sounds represented by the notes on a musical score are instants of time given values by the marks on the page, and thus, they are all like flashes of light that happen in the moment and affect us in terms of understanding independent of complex neural analysis. They are Blake's grains of sand in sound. 'To perform music you need only the phrase you are in... For the time he is in the music, provided it has his full attention, he forgets the abyss at his back. He has continuum. But when the music stops, he falls out of time all over again' (Wearing, 243). The key phrase here is, I think '*in* the music,' because actively listening to and making sound is a pres-ent-tense unfolding of experience that requires a domino-effect of cau-sality to occur, connecting moments of time as they pass us. In Victor Zuckerkandl's words, 'It is...a condition of hearing melody that the tone present at the moment should fill consciousness *entirely,* that *nothing* should be remembered, nothing except it or being beside it be present in consciousness' (Zuckerkandl in Sacks, 228).

The mystery of it all deepens again, however, when we seek to understand how structure and form—the stringing of a series of indi-vidual notes on a kind of 'washing line' of sonic time—can affect us emotionally. We may explain that experience and our ability to interpret music are a developing thing, given neural structures that are continuously learning and modifying themselves as new material is gathered and stored that 'our brains learn a kind of musical grammar that is specific to the music of our culture, just as we learn to speak the language of our culture' (Levitin, 108). Nevertheless, it remains remarkable that a response as immediate and involuntary as a heartfelt sob can be engendered by a seven-second sequence of notes in a Mozart piano concerto. Such a work as this is technically structured and exists within its perfection of form, but the patterns of sounds, sometimes so short while contained with the greater whole, provide instants of personal connection to a listener that will vary from individual to individual. Of these moments, Richard Strauss wrote that 'untrammelled by the mundane form, the Mozartian melody...hovers like Plato's Eros between mortality and immortality – set free from the Will – it is the deepest penetration of artistic fancy and of the subconscious into the innermost secrets, into the realm of the "prototypes"' (Strauss, 76).

The composer Jonathan Harvey has written that 'the element of mystery – a sense that something miraculous, beyond rational explanation, is taking place – is a crucial component of the experience of inspiration for most composers' (Harvey, 3). Harvey frequently used electronic sounds in his works, sometimes blending them with traditional instruments, and at others mixing with 'found' recordings from the physical world as in *Mortuous Plango, Vivos Voco*, where electronics comment on—and transform—recordings of the voice of his choir boy son with one of the great bells of Winchester Cathedral. There have been a number of occasions through musical history when composers have voiced frustration with the potential for existing man-made instruments to convey the subtle sounds heard in the imagination, keeping in some senses, the music locked within the physical. There is of course the ever-present metaphor of birdsong to turn to as a way of in every sense giving song flight, and Harvey is by no means the first to have explored this field to powerful effect. One of his later works, *Bird Concerto with Piano Song*, composed not long before his death in 2012, mixes real birdsong with electronics and orchestral and pianistic sound, offering a new grammar in which as Harvey himself wrote in a programme notes, 'real birdsong was to be stretched seamlessly all the way to human proportions – resulting in giant birds – so that a contact between worlds is made' (Harvey CD, 4).[3] The nature of the sound itself less linear than clustered and created through a series of *moments* in which the idea of birds in brilliant light seems to echo the flashes of the morning light of California, where the work was begun. As Arnold Whittall added in the programme note, 'the overriding aesthetic quality of the piece throughout is that of "playing with the idea of song": the basic contrast between organisms which sing and instruments which can (only) play fuels a transformational drama...' (ibid., 6).

Predating Harvey's work in this field is the literal notation and transfer of birdsong by the French composer, Olivier Messiaen. He had been experimenting with the idea of incorporating the purity of natural song since the 1930s, but it was a work created in the darkest of circumstances in which the first named use of birdsong in his work occurred. *Quatuor pour la fin du temps (Quartet for the End of Time)* was premiered in 1941. Messiaen wrote the piece while a prisoner of war in German captivity, and it was first premiered by his fellow prisoners. It has come to be recognised as one of his most important works. Messiaen was 31 years old when France entered the war. He was captured by the German army in June 1940 and imprisoned in what is now Zgorzelec, in Poland, but at the time was Görlitz, Germany. While in transit to the camp, Messiaen began developing a piece that would eventually become the *Quatuor.* The quartet was premiered at the camp to an audience of about 400 fellow prisoners and guards on 15 January 1941. The musicians had acquired what instruments they could—the cello was bought with donations from camp members. Although not named in the score, Messiaen cites in his commentary to the work, the blackbird and nightingale, and as Robert Sherlaw Johnson has pointed out, 'the characteristics of the different melodies are sufficiently well defined to make it clear that it is the violin which has the nightingale's song and the clarinet the blackbird's throughout' (Sherlaw Johnson in Hill, 251). The composer's own comment on this opening movement is itself a statement of great beauty:

> Between three and four in the morning, the awakening of birds, the dawn chorus: a solo blackbird or nightingale improvises, surrounded by a shimmer of sound, by a halo of trills lost very high in the trees. Transpose this onto a religious plane and you have the harmonious silence of Heaven. (Messiaen, programme note[4])

These magical and healing voices occur in the movement entitled, 'Liturgie de cristal.' Although the main subject theme of the *Quatuor* is a massive one, drawing on imagery and ideas contained in the Book of Revelation, yet the poignancy contained in these 43 bars of music speaks for what is left unsaid. Later works, such as the large-scale cycle for piano, *Catalogue d'oiseaux*, return specifically to the literal transcription of what Messiaen heard and which he wished to use to give a voice through a man-made instrument. In this great work, he sought to paint a portrait of particular birds individually, while at the same time giving an impression of their natural habitat and associating it with songs of other birds found within the same region. Thus, his interpretation of the Curlew is set within the bleakness of Ushant off the coast of Brittany. Also haunting the piece are the passing cries of Terns, gulls, a Guillemot, Oystercatcher and Turnstone. The mind's eye sees a vast expanse of water, as 'fog and night descend over the sea, the foghorn sounds its lugubrious note and – in the darkness – the last cries of the sea birds are heard. Finally the cries of the Curlew recede into the distance leaving only the cold night and the sound of the surf' (Sherlaw Johnson, album programme note[5]). Such music requires an extraordinary degree of poetic concentration by the performer (Sherlaw Johnson's 1973 Argo recording remains breathtaking in this respect), as well as deep listening and focus on the part of the listener. Yet nowhere, and at no time, would Messiaen encounter the degree of intensity of listening as shown by his prison audience on that day in the Nazi prison camp during the bleak January of 1941.

THE SOUND OF LISTENING

The key element in our relationship with an image—visual or aural—is the response that occurs within us, and the potential mistake we may make in our assessment of how that relationship operates lies in a failure to appreciate that seeing and listening are not passive, but active and creative acts. Gaston Bachelard draws our attention to Shelley's memorable passage from Act 4 of *Prometheus Unbound*, to underline the power of the dynamic imagination to interpret sound in terms of harmony:

> Listen too,
> How every pause is filled with under-notes,
> Clear, silver, icy, keen, awakening tones,

> Which pierce the sense, and live within the soul,
> As the sharp stars pierce winter's crystal air
> And gaze upon themselves within the sea. (Shelley, 302)

Bachelard, musing on this passage, comes to the conclusion that 'all space vibrates with the lively noises of the cold. There is no space without music since there is no expansion without space. Music is vibrating matter' (Bachelard, 49–50).

Just as a sound can transport the imagination across the globe, or even the universe, a musical phrase or a harmony has the potential to move the mind through a lifetime, or through centuries. Janet Cardiff's 2001 sound installation, *Forty-Part Motet*, is based on the English Tudor composer Thomas Tallis's great choral masterpiece, *Spem in Alium*. A visitor to the gallery in which the work is installed, in New York's Museum of Modern Art, will first remark on the technology, the forty speakers on stands, in a circle, singing to one another across a bare white room. There is elegance about it, a pleasing minimalism. Beyond that however, there is the sound of late sixteenth-century churches, sonic instruments in their own right, the great spaces in which this music once echoed and in which even today it finds its truest expression. It is music created in partnership with sacred architecture; the interweaving of the voices is seamless, reminding the listener of Tallis's mastery of the polyphony of his day, and Alessandro Striggio's great choral masterpiece, *Ecco si beato giorno*, which inspired this counterpart. The blend of the voices makes it impossible to hear the breathing of the singers. English Tudor polyphony and its European equivalents sought to create the music of angels, and angels (of course) not being mortal, have no need of breath.

Listening to Tallis's music—its notes as they pass, one by one, even in a recording—the mind is transported by association to the giant reverberating spaces for which it was conceived. I am taken too by the mind's suggestion from this room into other acoustics, and to recall other ancient music made for them, such as the soaring spaces of Renaissance Venice, where the architects Sansovino and Palladio worked in partnership with composers such as Adrian Willaert in the development of the polyphonic choral tradition and its relation to Place, specifically in that instance, in St Mark's Church. For anyone who has been fortunate enough to visit these places, the mind provides the reference of a mnemonic image which in turn informs the sound in another physical context but at the same time, internally. Listening, my sonic universe is expanding before my very ears. Cardiff's work is art in remarkable in that —while through sound it evokes the memory of architecture—simultaneously, it sends its own message from its inward-facing speakers, seeming as they do to discourse together, and it is the metaphor contained in this manifestation that in the end rings most strongly that of voices in harmony, be they contained within a room, or reaching us imaginatively and culturally from a sacred space half a world away. The presence of sound either as art itself, or as an idea suggested by it literally or tangentially in the mind, enhances the experience of our visually immediate creative environment and perhaps changes it while adding layers of meaning. As Picasso's *The Old Guitarist* (1903) confronted the accepted conventions of the art establishment of the time by flattening and fragmenting pictorial space, and as the Second Viennese school composers such as Schoenberg and Berg led music into new realms of sound in the early years of the twentieth century, so might we bring our personally created soundscapes to what we see, challenging ourselves by finding new and strange sonic worlds within worlds and like the poet Wallace Stevens and his 'Man with the Blue Guitar' come to understand that things may not quite as they are...or seem to be.

The Japanese composer Tōru Takemitsu fused tradition with innovation and, significantly, sound with silence in his music, the impressionistic minimalism of which shows a thread of influence from Debussy, but also Webern, Schoenberg, Cage and particularly Messiaen. An early interest was the Japanese garden in its colour and form; his *Spirit Garden for Orchestra*, commissioned by the Hida Furukawa International Music Festival, was inspired by its grounds, as a Shinto shrine. In a programme note to the work, Takemitsu wrote of the music as being based on three chords:

> These chords, accompanied by changes in tone colour...are an ever-present undercurrent, vibrating at the fundamental, from which a musical garden is composed. The "objects" of sound placed about the garden... change their forms through the changes in the angle of viewing which result from moving around the garden.[6]

This series of single sounds and clusters of melody form a work that is concentrated and subtle, celebrating the vibrations of air and matter as they interact across sacred ground.

In the next chapter, we will consider, among other things, the sound of birdsong itself, and in a sense, the closest we come as humans to the voices of birds is in the orality of the folk tradition. We might suggest that there is no sound without listening, without the conscious recognition of a sonic disturbance of the air. Even the cry of the last of a species, calling for a mate that no longer lives, is an audible voice and so may be heard by others for whom it is a sad irrelevance. Yet we may wish to keep in mind at the same time the statement of Barthes,' quoted earlier that there may be 'something inaudible' in music. Certainly, rhythm is a presence in much more than sound, while Emmanuel Levinas has suggested that 'musicality belongs to sound naturally' (Levinas/Hand, 133). Indeed 'the whole of our world, with its elementary and intellectually elaborated givens, can touch us musically...' (ibid., 134). Our presence is a key aspect of the partnership, because we alter a space by being present in it, and the intensity of anticipated sound can be almost palpable. The mystery of sound is that it links the physical with the ethereal, which is why the tolling of the bell is so charged with meaning, as a movement from sound, onwards into a silence that possesses an indeterminate and—of course—an inaudible beginning and end. It is as must be self-evident by now, a poetic means of expression in which the material sonic presence gives way to the imagination, as Brandon LaBelle has suggested: 'Sound forges a dynamic link between concrete experience and what seems immaterial...So might sound be understood as a vehicle, as something that configures a connection between private and public life?' (LaBelle in Belinfante and Kohlmaier, 73). To catch the temporal sound as it passes us, as the eye seeks to hold on to an image of flashing light on passing water, requires a very active form of attentiveness and aural commitment.

A community that possesses a capacity for active listening hears the world in a much more 'tuned' way than a social group that has become desensitised to sound through a consistent urban soundscape of over-amplified and synthetic sound. To make a sound, we break a silence, and to listen to a silence is to be aware of it as a presence. The poet and critic, Jeremy Hooker, writing the text that became his autobiographical sequence, *Under the Quarry Woods,* and looking across the valley from the study at his home in Treharris, South Wales, found his eye drawn to a Quaker burial ground. Silence is at the heart of Quaker worship, a fact that in turn led Hooker to consider, 'does the listening heart and mind leave a different silence?' (Hooker, 17). It is instructive to learn from the experience of a culture for whom listening is actively practised, as for example a society that has oral history and song at the root of its culture, and this may sometimes be most demonstrable within a community that is 'ring-fenced' geographically, thus preserving a degree of isolation and therefore a protection of its traditions enforced by its locality. For example, Tory is a small island in the Atlantic, just 3 miles long by 1.5 miles in breadth at its widest point, and 9 miles off the Donegal mainland in Ireland; such is its situation that it has a unique culture of song and story. In his monumental study of Tory Island's oral traditions, Lillis Ó Laoire has written of the 'vital necessity of listening' to this social group as it absorbs the songs that are such a vital part of its culture. 'Tory is a gaelic speaking community where the senses of sight and hearing are dominant factors in the learning process' (Ó Loaire, 89). Not only the absorption of material from oral sources, but also the complete accuracy of its presence in the memory, facilitating faithful reproduction when sung or played subsequently, was a crucial part of the preservation of the tradition:

> Hearing, or listening, to use the term signifying engagement, forms the foundation of...the song process...Listening provides the key for the morality and ethics of song among both those who seldom or never sing and those who are recognised singers. Performers are required to take on this responsibility and to continue and maintain it. (ibid., 89–90)

John Blacking has observed that 'in societies where music is not written down, informed and accurate listening is as important and as much a measure of musical ability as is performance, because it is the only means of ensuring continuity of the musical tradition' (Blacking, 10). This awareness and alertness to sound is a tuning that extends by its very nature beyond song into language and the surrounding natural world, so that it may be said that a community with an orally led culture hears the world's subtlest signals in a highly sophisticated way, just as a bare foot on earth 'reads' texture, temperature and terrain in a way that a shod foot never can. It is the musicality of language that is also preserved in this process, and the phonetic sound of a gaelic song in its original language cannot be replicated in a translation. For example, the first line of the Tory Island song, 'Young Donal': 'A Dhónaill Óig is tú pór na ngaiscíoch' possesses a sonic colour and richness of orality, even visible to the eye on the page, that a translation 'O young Donal, you're the seed of heroes' (Ó Loaire, 306–7) can only hint at, particularly when wedded to the melody of which it is itself an integral part.

I have written elsewhere about the concept of an internal sound that belongs to us and which is the germ of our sonic identity, something that joins with the sounds of the outer world in either harmony or discord.[7] It is something to which we come closest when we consider cultures, peoples and societies in which oral communication retains its power over the forces of media and the noisy world that so easily drown it out. In Ireland, Sean-nós singers such as Joe Heaney (1919–1984) demonstrate this intimate essence that lies within the heart of such cultural traditions. The intricacy and ornamentation of this particular style, traditional to the west Galway region of Ireland, was never gratuitous, but a way of lingering on a sound moment like the memory of a beloved location in the mind:

> There's some places you want to hold on to more than other places, that's when you put the grace notes in, you know. So that takes out the full depth and meaning of the song...
>   and you hold on to that particular place...I love it so much that I don't want to leave it. I just want to hold on to it as long as I can while I'm singing it. (Heaney, int. Cowdery, 1[8])

In an earlier interview, made by Ewan MacColl and Peggy Seeger at the MacColl's home in Beckenham, Kent during 1964, Heaney sought to explain the mysterious process of his music, almost as it were, to himself:

> You're just singing the song to yourself. And that's the way you'll find it, even tomorrow if you go back there. You'll hear the song, but you don't know where it's coming from half the time. I could sing a song for you now ten times tonight and each time I'd be different.[9]

There is, nonetheless, a sustaining essence that seems to know no boundaries, and transcends language:

> Even in Russian as I saw the other day. I could still feel it was there, more or less the same thing, although I couldn't understand a word he was saying, I knew right away. I even heard the Indians in Edinburgh and I could feel a similarity between them and the songs we have at home, although I couldn't understand a word of it. The same thing, the way was there, the style was there. (ibid.)

The meaning then is held in sound, like birdsong. It moves beyond language, and the individual sounds communicating mood, feeling and emotion in a way that even the singers themselves have difficulty in articulating by way of spoken explanation:

> I don't think the tone of voice has so much to do with it as the style it's sung in. That's not even said in words - it's in the voice there between the words and the language It's something in the voice that keeps on even though the words aren't spoken. It still keeps going. I think the best way to explain it is you'll keep the song alive, all the time even though you're not pronouncing the words. (ibid.)

Notwithstanding this, the song here is still a story, a communicable narrative, with or without words: 'They don't tell you to sing a song. *Say* a song…that means you're telling a story. And without the story, the song is lost' (Heaney, int. Cowdery). Joe Heaney—or to give him his Irish name, Seosamh Ó hÉanai—is the subject of *Song of Granite*, a film by Pat Collins, whose previous work, *Silence*, was such a subtle and profound comment on the need to find the receptacle of stillness into which we may place ourselves. Collins is deeply conscious of the small sounds that are so much a part of our personal music, and to which we give less outward expression than did our forebears:

> The writer Desmond Fennell said to me one time, that its the first time in 2,000 years of civilization that boys don't whistle and women don't sing doing the house work. Now it may sound a little chauvinistic but there is something in it. And if you ask a lot of young boys if they whistle, you find there is truth in the observation. My mother sang all the time. And if you think about boys or adults whistling, then you imagine that they must be feeling OK with the world. They could be announcing the fact that they are present – but they could be just, well care free! It's possibly a sound that we miss unknownst to ourselves. Is whistling not a positive affirmation of life?[10]

In *Silence*, Eoghan, the sound recordist in search of sound beyond human intervention, is exiled literally not only from his home, but also in a sense from himself, both his past self and himself in the present. 'The journey to record silence burns off in the film. At the start he is methodical, and then he begins to record by the sea, then later in the rain. None of this really makes sense if you're trying to record a silence. In the end he is not even recording; he is just listening – to what is around him and what is inside him' (ibid.). In this subtle film, there is a sense that only when Eoghan recovers the strength to sing, can he re-engage with his past.

The return to a still small voice inside is mirrored by the physical journey made through time and place, which is of course what every pilgrimage is at root, to travel through worldly space in order to find the sound of self, the defining song fragment that has faded until it becomes all but inaudible. In his great Icelandic epic novel, *Independent People*, Halldór Laxness expressed it thus:

> His mother taught him to sing. And when he had grown up and had listened to the world's song he felt that there could be no greater happiness than to return to her song. In her song dwelled the most precious and the most incomprehensible dreams of mankind. The heath grew into the heavens in those days. The song-birds of the air listened in wonder to this song, the most beautiful song in life. (Laxness, 226)

The Whisper at the Window

The wind in trees. Air made audible. The music the composers seek to emulate is present all around us. In the end, it comes down to this: the physical world of which we are a part plays like an orchestra constantly, and like an orchestra, it is made up of instruments of an enormous range of frequencies, volumes, tones, timbres and pitches. The positioning, height and proximity of the buildings with which we are surrounded can create wind tunnels and sounding boards; alleyways, railings, gutters can produce strange musical sounds when the wind travels through them or against them, and differing levels of noise and sonic intensity contrast with punctuating silences that may be themselves together create a form of 'found' music. Responding, albeit perhaps unconsciously over time, our very voices may echo our terrain: compare a New Yorker's tight, clipped accent with the wider stretch within the tones of a resident of San Francisco. We are wildlife within a landscape, and sometimes where we are and how much we travel govern the colours of our human music, just as migrating birds learn new habits through influence and adapting to place and circumstance. We will travel next to the fundamental source of it all: the movement of the atmosphere upon which it all depends. The sound of weather lies at the core of our sense of place; it has its own music, but the idea that it could be used to play on man-made instruments became a fascination in the late eighteenth and early nineteenth century. The concept of the Aeolian Harp, named for Aeolus, the Greek god of the wind, was one familiar in the ancient world, but was first described by Athanasius Kircher in 1673. During the Romantic era, it became popular as a household instrument, in essence a wooden box which included a sounding board with two bridges, across which were stretched strings either of the same or varying pitches. It would be placed usually on the sills of open windows, allowing the breeze to play across its surface. Thus, it made sound without human intervention and acquired a ghostly, mystical reputation; apart from the fact that it is the only stringed instrument to be activated solely by the strength of the wind, the Aeolian Harp is the only stringed instrument that plays solely harmonic frequencies. Unlike a wind chime, there is no percussive element to the harp, only harmonic frequencies played in sympathetic response to the winds, producing an ethereal music that some found almost supernatural. These qualities made it a popular subject for poetry, beginning in the mid-eighteenth century in 'The Castle of Indolence' by James Thompson. Shelley in 'Prometheus Unbound' evokes a form of celestial sound 'Kindling within the strings of the waved air/Aeolian modulations' (Shelley, 301). Likewise, Samuel Taylor Coleridge was fascinated by the instrument, and owned one to which he alluded in two notable poetic examples, 'The Eolian [sic.] Harp' (1796) and 'Dejection: An Ode' (1802).

For Coleridge, the sound affected him according to his mood at the time of writing; it could be seductive, even erotic:

> How by the desultory breeze caress'd,
> Like some coy maid half yielding to her lover,
> It pours such sweet upbraiding...' (Coleridge, 101)

On the other hand, it had the capacity to make a dark mood even darker:

> ...the dull sobbing draft, that moans and rakes
> Upon the strings of this Aeolian lute,
> Which better far were mute. (ibid., 363)

Notwithstanding, it was, above all:

A light in sound, a sound-like power in light,
Rhythm in all thought...
...Where the breeze warbles, and the mute still air
Is Music slumbering on her instrument. (ibid., 101)

In both poems, the sound of the instrument is governed by what we must assume were differing climactic conditions, those of a balmy evening, contrasted with a more turbulent night in which the wind becomes a 'mad lutenist' (ibid., 367) and echoes the screams and moans of tortured souls. Yet the sound is none of these things until the senses of the listener interpret them. This is not a music made from the consciousness—or even the unconsciousness of a human agency, but a sound upon which the mind places a meaning, dependent on the context of feelings and emotions at the time of hearing. 'Every sensory interaction relates back to us not the object/phenomenon perceived, but that object/phenomenon filtered, shaped and produced by the sense employed in its perception' (Voegelin, 3). One might add to that, the idea that all listening has a context based on mood, emotion, acoustic space and time. As Alexandra Harris has said, 'the mind...is the harp to be played on... Nature could only echo back his own mood, and thus could never change it...Our relationship with nature is a fine balance between what we give and what we receive' (Harris, 228–9).

The spaces between the notes in Silvestrov's *Bagatelles* assume greater significance, the more one concentrates on them; the ghosts that appear between the notes have time to materialise in the mind as their sounds reverberate through space, and the stone walls of the church acoustic add sympathetic resonances that charge the music as it spreads out from the keyboard, seemingly into infinity. At a certain point, one by one the man-made notes become owned by the space into which they travel, and we as listeners find ourselves in the midst of that space, each hearing our own personal music according to our state of mind. Taking one of the meanings of the word, 'arabesque,' that of an ornamental design consisting of intertwined flowing lines, we might apply it to the dance of words and musical notes, in partnership with the found sounds into which those words and notes fall. The ultra-sensitive person, touched by the phenomena of the world through highly tuned senses may suffer nervous problems and stress as a result. Yet such finely balanced sensory systems enable a complete engagement with the world of which less honed sensibilities are less aware, and perhaps more 'balanced' as a result.

Nevertheless, to tune to the minutiae of the sounds under sounds is to interpret and find meanings that enrich understanding of—and empathy with—the natural environment itself. Do we not speak of such sensibilities, after all, as 'highly strung'? Thus, we move from the creation of man-made sound in search of a natural root, the world of expression through which we have just progressed, onwards towards our subjective responses to the given sounds the world creates, although not specifically for us. It is appropriate therefore for us to begin this part of our journey, having considered the sound of the air, by turning our ears to the voices of the creatures that people it.

NOTES

1. Frumkis, Tatjana. 'Bagatelles and Serenades'—Album Booklet for *Bagatellan und Serenaden*, Silvestrov, Valentin. CD, Munich: ECM Records, 2007.
2. Street, Seán. *The Memory of Sound: Exploring the Sonic Past.* New York, Routledge, 2015.
3. Harvey, Jonathan. CD programme note: *Bird Concerto with Piano Song*, London: NMC Records NMC D177, 2011.
4. Messiaen, Olivier. Note taken from *Quatuor pour la Fin du Temps.* Berlin: Sony Classical 889853102, 2017.
5. Sherlaw Johnson, Robert. *Catalgue d'oiseaux.* London: Argo Records 2BBA 1005-7, 1973.
6. Takemitsu, Toru. *Spirit Garden: Orchestral Works.* Brilliant Classics, 8188.
7. Street, Seán. *Sound Poetics: Interaction and Personal Identity.* Palgrave, 2017.
8. Heaney, Joe/Cowdery, James R. *Say a Song.* Washington, DC: Northwest Archives Series NWARCD 001, 1996. Cowdery's recorded interview with Heaney, made in 1979–1981, are now part of the Joe Heaney Collection in the Ethnomusicology Archives at the University of Washington.
9. Heaney, Joe. Interviewed by Ewan MacColl and Peggy Seeger, 1964. Full transcripts of these conversations can be found at the website of *Musical Traditions: The Magazine of Traditional Music Throughout the World.* http://www.mustrad.org.uk/articles/heaney1.htm.
10. Collins, Pat. Correspondence with the author.

CHAPTER 4

Signals from Near and Far Shores: Voices from the Natural World

Abstract So to the sounds of the natural world, its flora and fauna. We must seek to place ourselves *within* the sonics of nature, rather than as outside observers. We understand that much of the sound and vibration with which creatures and plants communicate is beyond our ability to hear; new technologies may help us detect the minutiae of sound around us, but likewise other man-made creations can damage the natural world beyond repair. We discover bird dialect, and how migratory and non-migratory birdsong is affected by environment. We visit Henry David Thoreau beside Walden Pond. The sub-aqua world is beginning to reveal its many voices, and the earth itself—the shifting sands of its surface and the subterranean murmurings beneath the soil—asks us to listen.

Keywords Birdsong · Natural world · Sub-aqua sound · Sand

Songs of Love and Fear

An early evening in late June. An English garden. It is raining. A straight, heavy even rain falling through air virtually unmoved by any breeze. It is very still. Only the sound of the rain on the foliage. It is as though the weather has suppressed all other sounds. Yet not quite all. A single blackbird is singing, high in the tree over there on the right, at the corner of the field. The song is vivid, piercing, a total focus for the attention, now I know it is there. I think of 'Adlestrop', Edward Thomas's most famous poem, conceived at precisely this time of the year, although in different weather:

© The Author(s) 2019   63
S. Street, *Sound at the Edge of Perception*, Palgrave Studies in Sound,
https://doi.org/10.1007/978-981-13-1613-5_4

Yes. I remember Adlestrop—
The name, because one afternoon
Of heat the express-train drew up there
Unwontedly. It was late June.

The steam hissed. Someone cleared his throat.
No one left and no one came
On the bare platform. What I saw
Was Adlestrop—only the name

And willows, willow-herb, and grass,
And meadowsweet, and haycocks dry,
No whit less still and lonely fair
Than the high cloudlets in the sky.

And for that minute a blackbird sang
Close by, and round him, mistier,
Farther and farther, all the birds
Of Oxfordshire and Gloucestershire. (Thomas, 24)

Even before we read the poem, we are struck by the title, that strange village name, sounding somehow unreal, almost comical: a set of syllables that must have been the first thing to catch the attention of Edward Thomas, coming across it 'unwontedly.' Next, the study of sound under sound, or rather sound under apparent almost stasis and silence, because the mind first registers a near eventless vacuum that gradually fills with sonic activity, perspective and significance. The perfect poem of the growing awareness of sounds as all else in the world steps back and our emotional and psychic antenna tune in. It prints on the memory. Thomas, writing just before the First World War, encapsulates an English bucolic idyll in sixteen lines, and the peace of it all, the layers of gentle sound the moment holds, is all the more poignant because we know what happened next to the world. Here, the rain softly falls, and my blackbird sings on. I try to do what Thomas did: I listen beyond the bird for more distant sounds. I am not in a Cotswold village like Edward Thomas; I am in the suburbs of a great city in the north-west of England, listening through the rain. Yet it seems at this moment that the concentration of calm moist air and the single song is absolute. There is only this bird and me in the world, until, over the field and the houses beyond; there is the single hoot of a train waiting for a signal to change. Edward Thomas and I, both with our birds and our trains and our stillness, listening as the sounds in the landscape's perspective gradually reveal themselves. There is, as Italo Calvino says, 'the moment when the silence of the country-side is distilled in the ear cavity into a fine spray of sounds...The sounds contend with one another, hearing is able to constantly discern new ones, just as to fingers unwinding a ball of yarn every strand proves to be woven of ever finer and more insubstantial threads' (Calvino, 96).

Science tells me that sound is a sequence of pressure waves propagating through a compressible medium, and requires itself to move molecules in order to travel. These molecules vibrate or collide, passing on the energy in the sound until it reaches my ears. The closer the molecules to one another, the further the sound is able to travel. It is this that enables sound to move across greater distances through water than air. Fog is made up of small droplets of water: we know it affects what we see, but it also affects what we hear, because attenuation by sound waves in fog is a function of the frequency or pitch of those sound waves. It is not accidental that most foghorns have a low pitch, because the sound can travel greater distances. All of which makes me wonder why I am not hearing more sound, rather than less? Then I realise that the world is indeed still going on; the voice of a neighbour, a dog barking, a distant metallic crash from the container port on the river, and beyond that the drone of an aircraft, still far away, but beginning its descent into the low ambient murmur of the city. As it was in the art gallery tea room, where this journey began, it is all a question of tuning in. Just like Edward Thomas, I become conscious of ripples of sound created in a pool of apparent silence, and there, still seemingly at its centre, the single bird. The nature writer Richard Jefferies, whom Thomas admired greatly, wrote that 'it is in this quietness that the invisible becomes visible. The vacant field gradually grows full of living things. In the hedges unsuspected birds come to the surface of the green leaf to take breath' (Jefferies in White, x–xi). Jefferies wrote those words as part of an introduction to Gilbert White's classic book of detailed observation, *The Natural History of Selborne*. Turning to White's entry for 18 April 1768, I see that his attention was caught by a bird's song too:

> The grasshopper-lark began his sibilous note in my fields last Saturday. Nothing can be more amusing than the whisper of this little bird, which seems to be close by though at a hundred yards distance; and when close at your ear, is scarce louder than when a great way off. (White, 45)

My response to what I hear is subjective. The fact that I hear the bird and consider the apparent silence around it, gradually becoming aware of perspective, distance and gradated sounds from other sources, speaks as much about me, and my presence in the world, as it does about the blackbird in the rain. The bird is not at the centre of my sound world—I am.

We hear the presence of things as they interact with other things; 'Listen to the wind!' we may say, but in fact we are listening to the wind as air as it comes up against or interacts with objects such as windows, or gutters through the downpipe, or beats through the trees. It is matter giving wind its voice. So it is with rain:

> We hear the rain not through silent falling water but in the many translations delivered by objects that the rain encounters. Like any language, especially one with so much to pour out and so many waiting interpreters, the sky's linguistic foundations are expressed in an exuberance of form... (Haskell, 4)

In turn, climate and location change the acoustics of the world, just as an echoing church or a soundproofed studio alters the effect of voices and musical instruments. A bird singing on a city lamp post sounds differently to the same bird in the depths of woodland. These are the sounds we hear, but there are infinitely more sounds that are beyond us. When we consider frequencies of sound, our first thought may well be ultrasound, which inevitably brings to mind the sonic lives of bats, for so long a source of mystery and even fear. Since the 1940s, science has been able to detect the high-frequency sounds of insects and others, and subsequently, to understand something of what these sophisticated audio systems mean and the purposes they fulfil. The basis of echolocation—the great weapon in the armoury of the bat—is a form of self-communication. A signal is sent, bounces against an object and comes back. A bat's echolocation may emit signals from ten pulses per second to two hundred pulses per second, depending on the circumstances in which it finds itself. While these pulses are extremely short, they employ a form of frequency modulation, moving through high to lower pitches with extraordinary sophistication. 'The bat's auditory system is capable of detecting nearly infinitesimal differences between the times of the returning echoes of the various frequencies, thus perceiving the distance, size and speed of the object with uncanny accuracy' (Morton and Page, 42). Yet a bat's prey may well itself employ similar systems in order to evade destruction. Thus as we sit and enjoy the warm apparent peace of a summer garden, violent battles full of sound and fury akin to a *Star Wars* action sequence are being enacted all around us:

> A bat screaming in at 200 pulses a second is making a lot of noise, and many moths can hear the approaching creature long before it can pinpoint its prey. So the moth can take evasive action…Others have the capacity for emitting their own ultrasounds, closely matching the bat's frequency and thus, in a sense, jamming the bat's sonar. (ibid., 43)

To counter this, bats have the ability to change frequency, moving their pulses higher, thus as it were, coming in under the moth's radar. Creatures this sophisticated were always likely to cause humans to have, at least, some misgivings, and even dread and disgust, particularly because for centuries, ignorance was based on being deaf to the frequencies of such advanced communications networks. The tiny sounds we can actually hear are but a fraction of the story, so the rest we fill with superstition. There are few poets who have written odes to bats, other than D.H. Lawrence and he summed it all up in one line:

> Bats must be bats. (Lawrence, 'Man and Bat')

Birdsong on the other hand has inspired poets since there were words to describe things. There have even been attempts at creating written representations of it, but these can be little more than general mnemonics, without the sonic richness of the voice itself. It was Edward Marston, writing in *By Meadow and Stream: pleasant memories of pleasant places* in 1896, who suggested that the song of the yellowhammer might be translated into words: 'It is delightful to hear the yellowhammer's song—his only song: "A little bit of bread and no cheese"' (Marston, 38). Whether Marston coined this imitation, or whether it derives from an earlier source into which he was tapping, is unclear. What is undeniable is that we anthropomorphise, create meanings, and absorb and appropriate birdsong, as with so much else in the natural world, as though it was all there for us. 'Forget what this beauty might mean, and focus on how it makes us feel. We hear passion, love, exuberance, also melancholy and longing' (Rothenberg, 14). Birdsong unlocks the poet in us, but birds do not know about poetry, its self-conscious appropriation of sensation as expression of something else, and that is one reason they delight and awe us. They surround us, familiar and yet strange, close yet alien; we share their world, but not their understanding of it. In one of his last poignant essays, *Hours of Spring*, Richard Jefferies, a prose poet of nature if ever there was one, wrote movingly from his sick bed of the realisation of the natural world's indifference:

> Today through the window-pane I see a lark high up against the grey cloud, and hear his song. I cannot walk about and arrange with the buds and gorse-bloom; how does he know it is the time for him to sing? Without my book and pencil and observing eye, how does he understand that the hour has come?... For they were so much to me, I had come to feel that I was as much in return to them. The old, old error: I love the earth, therefore the earth loves me – I am her child – I am Man, the favoured of all creatures. I am the centre, and all for me was made...Today I have to listen to the lark's song – not out of doors with him, but through the window-pane, and the bullfinch carries the rootlet fibre to his nest without me. They manage without me very well. (Jefferies, 21–4)

The 'oneness' yet 'separateness' of the fauna with which we cohabit is part of the mystery that writers and musicians have fed off in their work for centuries. In the previous chapter, we remarked on the works with birdsong that were key parts of the oeuvre of composers such as Olivier Messiaen and Jonathan Harvey. To them we must add the writing of Finland's Einojuhani Rautavaara, notably for his *Cantus Arcticus, Concerto for birds and orchestra* of 1972, which employed recordings of bird sounds made in the Arctic Circle and the marshlands of Liminka. The work rings with the calls of bog birds, shore larks and migrating swans, and it is music in which orchestral sounds blend with the natural world, creating a feeling that is typical in this composer's music, that of interaction between individuals and a collective. Ours is indeed a shared world. Yet we delude ourselves when we suggest that the natural world is something other than where we are, because we are not privileged observers but participants, albeit self-conscious participants who by chance have the capacity and ability to articulate our responses, to interpret, and—frequently—to misinterpret. In his remarkable book, *Being a Beast*, in which he chronicles his attempts at placing himself as far as possible within the exact experience of wildlife, Charles Foster asked—and sought to answer—the key question of kinship with nature:

What's an animal? It's a rolling conversation with the land from which it comes and of which it exists. What's a human? It's a rolling conversation with the land from which it comes and of which it consists – but a more stilted, stuttering conversation than that of most wild animals. The conversations can become stories and acquire the shape and taste of personality. Then they become the sort of animals we celebrate, and the sort of people we want to sit next to at dinner. (Foster, 20)

There, in a nutshell, is the compromise; we trade the ability to experience a reality for the capacity to express and interpret as best we can, that experience. In this, sound is both the key and the door the key will not open. In the last two hundred years, humankind has created artificial sounds in daily life that have exceeded those that issue from the natural world; prior to that, the sonic signals came more clearly and there was a knowledge of how to read them that was instinctive. When Mozart wrote *presto*, the fastest sound-generator in his world would have been the galloping horse, the fastest object above him, a bird. The loudest sound for him might have come from the farmyard: the call of a cockerel can reach a deafening 143 decibels, so loud that when in full 'song,' a quarter of the bird's ear canal closes to protect its own hearing. At the other end of the scale, observe the blackbird on the lawn, listening for subterranean activity. The hearing range of birds is most sensitive between 1 kHz and 4 kHz. The most commonly stated range of hearing in a human is 20 Hz–20 kHz, with the greatest sensitivity in the range 2000–5000 Hz. Thus bird hearing encompasses a narrower range of frequencies than human hearing but the main focus of their audio world is in the realms of pitch, tone and rhythm changes. This sense of relativity is useful, because scale is everything when considering—or reconsidering—the natural sound world.

We may associate echolocation with bats or seagoing mammals, but there are some birds that utilise the same skill, as in Oilbirds (*Steatornis caripensis*), one of only a handful of birds with the ability to echolocate. These nocturnal birds roost in caves across the tropical forests of north-western South America and spend a considerable amount of their time in the dark. In conditions where eyesight is irrelevant, individuals use sequences of clicks to build up a 3D image of their surroundings. Birds may use low-frequency sound location while migrating, 'by knowing the sounds of the earth as it is played upon by the weather – listening to how the winds play over mountain ridges like huge flutes and how the waves play the shores in their deep rhythms, slowly modulating the pressures of the earth's atmosphere' (Stocker, 132). Migrating flocks may fly too high for us to hear their group communications, but as they come in to land, we may hear their chirps and call notes, identification signals to one another that declare their presence within a specific space. Watching a murmuration of starlings is always an extraordinary experience, and the precision of these patterns has sound as a key ingredient. I have seen a flock of thousands of these birds settle within seconds in a disused building at nightfall, inch-perfect and without a hint of collision. The sophistication of birds' use of sound can be extraordinary and is almost infinitely variable, according to species. What sonically separates us most from the avian world—indeed from much of the natural world at large—is time. Charles Foster has explained this with sardonic humour:

> If, like many birds, you can hear sounds separated by less than two millions of a second, you'll know the baroque complexity of apparently bland birdsong. If you're a human hearing that, you'll fall to your knees…Only those blind to the velvet flow of a caterpillar's legs and deaf to the grunt of a crocus as it noses out of the earth don't worship, and often they can't be blamed. (Foster, 189–90)

We live our sound lives through differing pitches and frequencies. In order to come close to understanding how bats and birds such as swifts, swallows and starling negotiate the air, we need to slow the tape down. 'The acutely discriminating bird hears what I'd hear if I turned the speed of the birdsong right down. I can probably hear two sounds as distinct if they're around two hundredths of a second apart. The bird's getting in one second what it would take me about 2 3/4 hours to hear' (ibid.). There are bird sounds that we can interpret with little difficulty, and mostly these are alarm calls; the scolding of a blackbird, or the shouting of magpies. Birds are aware of their own 'family' and they are aware of objects of danger, and because we too, as animals, are primed to grow alert in times of risk, we can read their cries of fear and warning more than any other sound they make. As to the actual sound, and its interpretation and imitation, it can be problematic. As we noted in the previous chapter, composers such as Harvey and Messiaen among modern musicians have sought to represent the sound, and through the long history of music, there have been birds as both source of inspiration and content. On the printed page, it becomes even harder to evoke birdsong, all of which supports our feeling of wonder at it. John Clare surmounted the problem of sound on the page by actually creating words in situations when nothing in the lexicon offered itself as appropriate. For example, in 'The Flight of Birds':

> Whizz goes the pewit o'er the ploughman's team,
> With many a whew and whirl and sudden scream...
>
> ...From bush to bush slow swees the screaming jay,
> With one harsh note of pleasure all the day. (Clare II, 239)

In prose, there can be few better depictions of bird sound and bird presence than J.A. Baker's visceral book, *The Peregrine*. Baker's compulsive, obsessive observing of avian life—and particularly death—in the Essex countryside mostly to the east of Chelmsford is almost literally twittering with sound, and his writing is of animalistic intensity. To analyse it is like unpicking the techniques within a musical score: 'Grammatically, his prose is dense in metaphors, similes, verbs and adverbs; accentually, it is thick with stressed syllables. This combination of axe-knock stresses and ultra-kinetic syntax results in a style that is shocking to read' (Macfarlane in Baker, 194). It is almost like listening to a recording at times, so unique is the way in which Baker hurls the immediacy of the moment— physical and emotional—onto the page, and so often there is a sense of danger, of something about to happen, the strike of the Peregrine itself. Some random phrases will give a flavour: 'Crows in the elms are cursing and bobbing. Jackdaws cackle up from the hill, scatter, spiral away, till they are far out and silent in blue depths of sky' (Baker, 49). 'A moorhen called, and tinkling goldfinches hid silently in thistles' (ibid., 52). '...the far bugling of a curlew's desolate cry echoing through the harsh staccato chatter of a hawk' (ibid., 53). Baker writes as birds cry, and there are times in the book where the words seem to emerge out of the bleakness of wild places like the creatures that inhabit them:

> Redshanks shrill and vehement; never still, never silent. The faint, insistence sadness of grey plover calling. Turnstone and dunlin rising. Twenty greenshank calling, flying high; grey and white as gulls, as sky. Bar-tailed godwits flying with curlew, with knot, with plover; seldom alone, seldom

settling; snuffling eccentrics; long-nosed, loud-calling sea-rejoicers; their call a snorting, sneezing, mewing, spitting bark. Their thin upcurved bills turn, their heads turn, their shoulders and bodies turn, their wings waggle. They flourish their rococo flight above the surging water. (Baker, 60)

We shall have cause to return to both Clare's and Baker's avian words later in this book. The nature of a bird's voice may be affected by where it lives and how mobile it is outside of its native terrain. In other words, birds may sound differently, according to their travelling habits. Studies by Donald Kroodsma in the USA have explored variations between home-loving birds and those that migrate, as well as regional differences: in other words, dialects. Based in Seattle, he observed that 'the Bewick's wrens are resident, or non-migratory; a male learns his songs after dispersal and then stays in his neighbourhood for life. The young wrens enter the song neighbourhood in the same way that young song sparrows do, and the songs of the father appear to be forgotten' (Kroodsma, 67). On the subject of sparrows, Kroodsma noted that there was a distinct difference between the sounds of birds from the east and west of the continent: 'Western birds are resident, staying on their adult territories in the same neighbourhood for life; the natural selection apparently favours shared, learned songs in these circumstances' (ibid.). It is very much analogous to human dialect, and one might even extend the idea into locally based folk song. Migrating birds, on the other hand, become exposed to different circumstances, and thus their voices become affected:

> Impose some tough weather on these birds and make them migrate…and the microdialects break down….The extreme case is the sedge wren, which migrates not only between seasons but also *within* seasons [my italics] so that neighbours know each other only briefly; the sedge wrens' style of song development is also extreme, with males no longer imitating songs but making them up instead. (ibid., 67–8)

Even within the relatively small compass of the British Isles, common garden birds may display regional dialects; the British Library contains numerous recordings of UK birdsong, and to the trained ear, variations from different parts of the country show themselves in subtle ways. Just as the human voice is recognisably of our own species, yet displays its geographical origins in language and accent, so the sounds, for instance, of a chaffinch from Cumbria, Northamptonshire or Hampshire contain sounds that enable ornithologists to place them geographically. The late Jeffrey Boswell explored the question of birdsong, and how instinctive it is, his question being, is it instinctive or learned? The experiment to explore this question included the hand-rearing of chaffinch chicks in a Cambridge laboratory. The project involved taking newly laid eggs, incubating them separately in soundproof chambers, then hand-rearing each bird in individual and acoustic isolation, and then analysing the sounds each bird produced as it grew. This was inevitably controversial, but the results of this quest into the idea of 'nature versus nurture' were interesting, when, after a year, the young birds sang their first recognisable songs.

> In the wild, a young bird would add the finer details during the first few weeks of its life, having learnt them from its father and other cock chaffinches within hearing; and again in the following year when the wild chaffinch's first breeding season approached, it came to sing in competition with neighbouring territory-holders. These recordings of the impoverished and the full songs illustrate the difference, exposing what is apparently the embroidery that has to be learnt...[1]

The British Library's collection of birdsong is an invaluable resource and owes much to the donations of benefactors who have over the years shared the results of research and in particular recordings. Among these is the work of William Homan Thorpe (1902–1986) who was Professor of Animal Ethology at the University of Cambridge, and highly significant British zoologist and ornithologist. Together with others, including Nikolaas Tinbergen, Patrick Bateson and Robert Hinde, Thorpe was a major contributor to the growth and recognition of behavioural biology, and during the 1940s he pioneered the use of sound spectrography to analyse birdsong in detail, at the time using the only example of the technology available in Britain. A spectrograph is an instrument normally associated with light that is used to separate and measure the wavelengths present in electromagnetic radiation and to measure the relative amounts of radiation at each wavelength. In other words, in its most familiar function, it obtains and records the spectral content of light or its 'spectrum.' The spectrograph splits or disperses the light from an object into its component wavelengths so that it can be recorded and then analysed. In terms of a sound that changes in time, such as a spoken word or a bird call, the device enables a more complete description by examining how the Fourier spectrum changes with time. The term *Fourier transform* refers to both the frequency domain representation and the mathematical operation that associates the frequency domain representation to a function of time. In a graph called the sound spectrograph, frequency of the complex sound is plotted versus time, with the more intense frequency components shown in the third dimension or more simply as a darker point on a two-dimensional graph. It was this level of detail that enabled the identification of specific details of regional variation in the song of birds such as the chaffinch.

On Thoreau's Front Porch

The world comes to us through all our senses. Sound is a part of a communications system of such complexity that at any one time the brain is processing signals beyond our conscious ability to register them, and yet they lodge themselves within the mind to establish a library of information and impressions of infinite subtlety and nuance. I will return to the theme of connection between visual art and its murmur of sound later in the book; the link is crucial because we are concerned here with the moment, and its impact upon us. A painting or a photograph has its origin in the preservation of the moment, its recording and freezing, the capacity to save an instant of sensation for all time. The sound we 'hear' when we look at a work of art is imaginative. The gallery in which I viewed the Courbet painting is relatively silent, but possessed of an acoustic in which small human sounds of movement and distant traffic noise outside become significant because there is nothing to detract from their intrusion...until the imagination is caught by the picture, and its window opens and admits our mind into its world, and we experience it through a process of interaction that involves us through emotional response memory and emotion. Sound is so often a recalled thing, and silence is not an absence but a receptacle. We have saved the experience of place, people, climate, weather and environment, and the stimulus of an emotive, moving or provocative object such as a work of art triggers a whole network of emotional echoes.

The small fragile sounds of ideas can be stillborn and can be drowned out by the noise of practicality and utility. At such key moments only the patience and silence of the notebook or sketch pad will do.

There is a poignancy in the instant, its sound and sensation, and the desire—even the need—in us, to hold on to it before the river of Time sweeps it away. Record by all means; a pocket digital recorder can play back a fragment of life and show you things you had not noticed before. At the same time, by writing a word, the very act of it, engages with the feeling of the moment. An artist, sketching a scene, commits such attention to the object in question that their concentration burns that moment into memory. Sometimes even the computer and its keys can seem too crude and non-human to catch the whispers of birth as a poem or a song or a drawing that will become a painting take their first faltering steps. There are times when only paper will listen.

At the other end of the process, at the point of consumption, that paper can also speak. When we read, we see and we hear through the window of the text by the same process that was begun when we looked at the picture. Passion persuades, and to enter the world of ideas created in a great book, such as Henry David Thoreau's masterpiece, *Walden*, is to hear a place talking through the words that so eloquently recorded it. Thoreau (1817–1862), essayist and philosopher among many other things, published his great book—a reflection on simple living in natural surroundings—in 1854, and it chronicles two years, two months and two days of living in a cabin he built by Walden Pond, on land owned by his friend, Ralph Waldo Emerson, near Concord, Massachusetts. His observations—sometimes scathing in relation to the growing noise of commercialism he saw around him in mid-nineteenth-century America— capture his developing thought in the context of immediate impressions of his surroundings. For Thoreau, the written word is 'the choicest of relics' and is 'the work of art nearest to life itself. It may be translated into every language, and not only read but actually breathed from all human lips; – not be represented on canvas or in marble only, but be carved out of the breath of life itself' (Thoreau, 96). We may imagine ourselves sitting beside him in the evening light, by Walden Pond, hearing the perspectives of the sonic world coming to us at varying distances, framed by the stillness of our immediate surroundings. Thoreau noted that 'this small lake was of most value as a neighbour in the intervals of a gentle rain storm in August, when, both air and water being perfectly still, but the sky overcast, mid-afternoon had all the serenity of evening, and the wood-thrush sang around, and was heard from shore to shore' (ibid., 81).

About a third of the way into his book, Thoreau devotes a chapter specifically to the sounds around him. It is one of the most magical things he ever wrote; almost half of the chapter is devoted to the sounds he hears coming across the trees from the railroad tracks, and we sense their intrusion into his earthly paradise. To him, these sounds mean commerce, the unnecessary bustle and hurry of modern life, industry, the desire to move, to make money: in short all the trappings of what he saw as the negative side of progress. Then, quite abruptly, it becomes quiet, and we hear with him, the sounds of nature as they flood back into the consciousness. One Sunday, he becomes aware of church bells, their sound coming from various sources in surrounding communities—Lincoln, Acton, Bedford or Concord itself. He notices that, when the breeze is propitious, these sounds become part of the natural world; this is not an intrusion, but a blending. A bell, after all, fades through the life of a single sounding towards silence, but the moment of inaudibility, where it melds with the ambient sounds around it, is subtle and elusive. Thoreau captures the idea perfectly:

> At a sufficient distance over the woods this sound acquires a certain vibratory hum, as if the pine needles in the horizon were the strings of a harp which it swept. All sound heard at the greatest possible distance produces one and the same effect, a vibration of the universal lyre, just as the intervening atmosphere makes a distant ridge of earth interesting to our eyes by the azure tint it imparts to it. There came to me in this case a melody which the air had strained, and which had conversed with every leaf and needle of the wood, that portion of the sound which the elements had taken up and modulated and echoed from vale to vale. The echo is, to some extent, an original sound, and therein is the magic and charm of it. It is not merely a repetition of what was worth repeating in the bell, but partly the voice of the wood... (ibid., 115)

A bell has many moods and many meanings. It interacts with us in mysterious ways and lodges itself uniquely in the mind. It travels through air as it dies, on its journey mixing with the world that listens to its message.

ULTRAMARINE

I have a vivid recollection of attending a lecture by the sound recordist Chris Watson at the Purcell Room on London's South Bank, in which he played a recording of the call of a Blue Whale, the largest creature that has ever lived on earth. Coming to the audience through a high-quality speaker system, in a room with sophisticated acoustics, designed as a platform to show off the nuances of intricate chamber music, the effect was devastating. Very low, a vibration almost more than it was sound, it consumed the room; the ear pulsed, the chest shook with it. This was a sound never intended to be heard through the medium of air: a sound designed to communicate across thousands of miles of ocean, through deep salt water. The sound of the Cetacean species, among them whales, dolphins and porpoises, was a part of myth and seagoing fable for centuries, sound systems so advanced that it has only been with the development of human devices to explore marine life through sonar and other tools that we are beginning to understand them, and even now they remain sounds—literally—from another world, full of awe and wonder. A detailed exploration of sub-aqua acoustics is outside the scope of this short book, and the interested reader is directed to the bibliography, and to other more specialised works in this complex and fascinating field. Suffice it to say that in auditory terms, the human species is an infant compared to many of its neighbours:

> Bat and dolphin use sound to survive and have greatly superior capabilities to current technology with regard to resolution, object identification and material characterisation. Some bats [for example] can resolve some acoustic pulses thousands of times more efficiently than current technology. (Assous et al. in Lin, 2015, 1)

When we turn our attention specifically to the life that exists below the waves, particularly the communication systems employed by Cetaceans, it becomes clear how much more we have to learn, and how much we do not understand or even hear, and perhaps how much we could enhance our existence were we ever to attain such a capacity.

> Dolphins are capable of discriminating different materials based on acoustic energy,...significantly out-performing current detection systems. Not only are these animals supreme in their detection and discrimination capabilities, they also demonstrate excellent acoustic focusing characteristics – both in transmission and reception....Whilst some elements of animal systems have been applied successfully in engineering systems, the latter have come nowhere near the capabilities of the natural world. (ibid.)

It is curious to remember that in 1953, Jacques Cousteau, the great underwater explorer, who to a generation of young people communicated the wonders of the multifarious life below the waves of our seas through a series of evocative films and television programme, could title one of his most famous books, *The Silent World*. Today, we are highly aware of the sounds within the sea, and technology is developing constantly through more and more sophisticated hydrophones and measuring devices. At the same time, it is shocking to understand how much other technologies pollute the world's waters with sound. Sound moves at a much faster speed in water than in air. The distance sound travels is mainly dependent on variables such as pressure and temperature, governed frequently by the depth of the water itself, as characterised by the thermocline layer, a region—occurring at differing depths in various parts of the world—below which while temperature remains constant, pressure continues to increase, causing the speed of sound to increase, while refracting the sound upwards through what are known as sound channels. This channelling allows sound waves to travel many thousands of miles without the signal losing very much strength. Above the thermocline layer, sound refracts downwards and when it comes to the bottom of the layer, it reaches its minimum. Thus, situation and circumstance govern the distance sound may travel underwater. The relative salinity of water is also a factor in sound velocity, and salt content may also be affected by such issues as evaporation. In sea water, sound can travel up to 33 metres per second faster than in freshwater partly because there are more molecules, in particular salt molecules, with which sound waves are able to interact. Thus, we become aware that this so-called silent world is full of sound that for most of us remains unacknowledged for much of our lives. The sound of colonies of snapping shrimp may not be familiar to many, but it may be the commonest sound on the planet.

> Snapping shrimp...use bubbles to help make their sound – sometimes for communication, but at other times to kill their prey...the shrimp closes its claws very rapidly, with the tips moving at 70 kilometres per hour, creating a jet of fast-moving water...The pressure drops in the rapidly moving water, low enough for the water to start boiling at sea temperature. A bubble of water vapour forms, which immediately collapses and creates a shock wave that stuns or kills prey. (Cox, 94–5)

A sound that goes on unheard by millions of other species, yet this constant activity in the shrimp's sub-aqua killing fields disrupts marine recordings and even, during the Second World War, interfered with submarine surveillance. This is only one example of the hidden noise within water, and scientists in the field of underwater acoustics are creating listening devices of constantly increasing sophistication, already capable of enabling us to hear what we otherwise might have considered to be silent creatures, such as feeding sea urchins, rasping and scraping algae from rocks deep below the ocean's surface.

All of this is important for us, if we are to understand the importance of sound to the life of fish and water mammals. It is also salutary to come to the logical conclusion that, if this is the case, then the more disruptive noise that interferes with this communications system can wreak havoc with marine life. A whale may call to a distant pod from many miles away, but the increasing clutter of man-made marine audio activity, intended and unintended, disrupts the signal just as much as crowded wavebands on a short wave radio. The noise of such things as oil rigs as they drill can be catastrophic to underwater life for many miles; shipping may appear relatively quiet to those of us watching from the shore, but below the waves a large tanker at full speed may sound deafening in relative terms, and this human-made cacophony can seriously damage the social networks of whales, and actually adversely affect their survival and reproduction. It may be sonic interference with sophisticated communications systems that creates incidents in which whole pods of whales and other cetaceans strand themselves and die on beaches in various parts of the world. Given the technologies for listening to marine life, still developing, with sensitive application, we may be able to reduce sound pollution and alleviate some of its more shocking repercussions. In the meantime, however, if this is mankind's revenge on the Snapping Shrimp, then it is a terrible one.

THE SONG OF THE EARTH

Soil, like water, conducts sound efficiently, due partly to its density; as we know, it is also highly responsive to vibration. When we consider how we feel the rumble of a subway train passing under us, or even the sound of footsteps as they pass over different terrain, we remind ourselves of the potential of the earth beneath our feet to be a major conductor of sound. Frequency is all in this environment, and the more intimate contact a species has with the earth, the more sophisticated is its interactive capacity with its location. Snakes for example lack both outer ears and eardrums, yet their jawbones act as coupling elements to pick up ground-borne vibrations, and deliver acoustic information to a cochlea-like sensory system. As sound travels readily and far in a dense substrate like soil, the snake's direct coupling to this substrate is highly efficient and enables the acquisition of information from distant sound sources.

Communication is reliant on many sensory criteria, and when we examine the sonic world of our earthbound neighbours, we must widen our concept of what we consider to be auditory in the traditional sense. The field of plant bioacoustics is a developing one; we might consider that shrubs, flowers and trees relate to the audio world only insofar as they are passive responders to the elemental forces around them, or perhaps at most, there is a measurable minute signal that is detectable as fluids such as water and sap travel through their structures. Yet we ourselves absorb signals from the world around us only according to our limited reception capability, as Marcus Anhäuser of the Max Institute for Chemical Ecology has written:

> There are dramas playing out in meadows, forests and hedges – dramas of death and downfall, hunting and enticement, attack and defense. Just because humans lack the appropriate senses, we fail to notice the daily fight for survival that continuously goes on between plants and their adversaries: insects, fungi, bacteria and viruses. Alluring calls, warning calls, calls for help – we rarely perceive them because they are not acoustically or optically encoded. We know them only as the seductive fragrance of a spray of flowers, or as the heavy bouquet of a freshly mown field. If we could hear them, the screams of battle would destroy any summer idyll. But people have no natural sense or comprehension of the molecular language of plants. (Anhäuser)[2]

The work of scientists such as Monica Gagliano of the University of Western Australia and others is beginning to make inroads into this fascinating area. Gagliano and her colleagues collected grain seeds from the field and exposed them to a range of laboratory experiments; It was found that whenever these seedlings' roots were subjected to a specific sound frequency—in this instance, 220 Hz—their orientation shifted in the direction of the sound, as though they were responding to signals received on minute antennae. In other words, they were, in effect, listening:

> In plants, both emission and detection of sound may be adaptive. While receptor mechanisms in plants are still to be identified, there is early, yet tantalising, evidence about plants' ability to detect vibrations and exhibit

a frequency-selective sensitivity that generate behavioural modifications...
We are growing increasingly doubtful of the idea that *all* acoustic emis-
sions by plants are the mere result of the abrupt release of tension in the
water-transport system. (Gagliano et al.)[3]

This implies an astonishing opening into the sonic environment around
us and reminds us powerfully that whether or not we hear these voices,
they are, nonetheless present all around us. 'As a successful example of
interdisciplinary partnership, chemical ecology has greatly advanced
our understanding of plants by unveiling their strikingly "talkative"
nature and the eloquent diversity of their volatile vocabulary' (ibid.).
Technology takes us to places beyond the capabilities of the naked ear,
just as radio telescopes open the skies and microscopes take us to the
meanings that lie invisible around us. It is salutary and humbling to
understand how complex is the sound world in its minutiae, and how
ignorant we would be of it, without artificial aids. In the field of archae-
ology, scientists have increasingly used such devices as ground-penetrat-
ing radar and seismic exploration tools to enhance discovery. At the turn
of the century, researchers from the University of Illinois developed a
high-resolution imaging system based on sound waves in order to detect
tiny objects buried in soil. It is a technology that has also been used in an
adapted form to detect landmines, seismic exploration by which a charge
is ignited and receivers read the reflected sound waves; the difference in
this instance is that the frequency used is higher, and therefore the reso-
lution is much greater. Here then is an example of instruments that enable us to 'see' through opaque substances using sound.

The world's transmissions are all around us, and the human ear,
when fully functional, is itself an extraordinary instrument of detec-
tion. The crime novelist Josephine Tey was a great observer of sound
as a key element of plot in many of her books. Her 1952 story, *The
Singing Sands*, is a clear example of elemental forces underpinning the
unfolding of a highly complex interweaving of clues. Early in the book,
her sleuth, Alan Grant, examines a scribbled text hastily scrawled on a
newspaper:

The beasts that talk,
The streams that stand,
The stones that walk,
The singing sand... (Tey, 10)

Grant muses on the meaning of these enigmatic words, the euphony and alliteration, the assonance and the uncanny puzzle they seem to offer the mind: 'Singing sand. Surely there actually were singing sands somewhere? I had a vaguely familiar sound. Singing sands. They cried out as you walked. Or the wind did it, or something' (ibid., 13). Grant's hunch is right; the dust of the planet is far from sonically inactive. There are many instances in which the song of the earth can be heard with the unaided human ear, and sand can be among the most vocal. Charles Darwin, writing in May, 1835, in *The Voyage of the Beagle* recorded stories of strange sounds originating from the landscape near the town of Copiapó:

> While staying in the town, I heard an account from several inhabitants of a hill in the neighbourhood which they call "El Bramador", the roarer or bellower...As far as I understand, the hill was covered by sand, and the noise was produced only when people, by ascending it, put the sand in motion. Upon reading an article in the *Edinburgh Journal*, I was surprised to find the same circumstances, described in detail on the authority of Seetzen and Ehrenbergh as to the cause of the sounds, which have been heard by many travelers on Mount Sinai near the Red Sea. One person with whom I conversed had himself heard the noise; he described it as very surprising; and he distinctly stated, that although he could not understand how it was effected, yet it was necessary to set the sand rolling down the acclivity. I can vouch for the quantity of loose sand lying on the bare granite mountains in the neighbourhood. From the position of the hill, and from the account I received, the phenomenon certainly does not appear to have any direct connection with volcanic causes. I may remark that a horse walking over dry and coarse sand, causes a peculiar chirping noise from the friction of the particles: a fact which I have several times noticed on the coast of Brazil. (Darwin, 311–2)

Similar phenomena have been noted around the world; on Prince Edward Island in Eastern Canada's Maritime Provinces, off New Brunswick and Nova Scotia in the Gulf of St Lawrence, lies 'Singing Sands Beach' at Basin Head Provincial Park. It has become famous, and a favourite tourist destination because of its almost pure white sand, but also because the sand itself, with a high preponderance of well rounded and highly spherical grains containing quartz, produces a squeaking or 'singing' sound underfoot. Many other examples exist; singing sand dunes at Kelso in California, the 'Booming Sands' in the Namib Desert in Africa, Porth Oer near Aberdaron in North Wales, 'Barking Sands' in Hawai'i and many others. Singing sands have been reported on 33 beaches in the British Isles alone, and the cause is usually the same, that is to say, it is caused by walking on the soft, white sand and producing a sonic interaction. Hugh Miller writes in his book *The Voyage of the Betsey*, published in 1858, of a summer holiday in the Hebrides, including this account of the 'musical sands' of Camas Sgiotaig on the Isle of Eigg:

> I struck it obliquely with my foot, where the surface lay dry and incoherent in the sun, and the sound elicited was a shrill sonorous note, somewhat resembling that produced by a waxed thread, when tightened between the teeth and the hand, and tipped by the nail of the forefinger. I walked over it, striking it obliquely at each step, and with every blow the shrill note was repeated. My companions joined me; and we performed a concert, in which, if we could boast of but little variety in the tones produced, we might at least challenge all Europe for an instrument of the kind which produced them. (Miller, 58)

By moving on, Miller and his companions identified the origin of the phenomenon. Passing over drier tracts, he describes an incessant *woo, woo, woo* rising from the surface, finding on inspection 'that where a damp semi-coherent stratum lay at the depth of three or four inches beneath, and all was dry and incoherent above, the tones were loudest and sharpest, and most easily evoked by the foot' (ibid.). Not all such sounds are caused by human contact, however; in cases such as the Kelso and Eureka Dunes in California, and other desert terrains, the sound is caused by wind blowing over the sand and setting up a reaction. Theories include the concept that the frequency of vibration is related to the thickness of the layer of sand, and its aridity. It would seem that only sand of a certain consistency—usually between 0.1 and 0.5 mm in diameter, containing silica and at a specific humidity—will have sonic potential. It is thought that in the case of dunes, sound waves bounce between the dry surface sand and the moister layer beneath, creating a resonance, possibly generated by friction between the grains, or air as it is compressed between them. The wind's interaction with sand is of course critical; indeed, scientists talk of 'Aeolian sand-ripples,' evoking the strange sounds reaching Coleridge across his windowsill in the last chapter. This factor, taken into account with sand grains of varying consistencies interacting with one another, plays key roles together in making the sound that comes with movement. Vaughan Cornish made an exhaustive study of sand movements in his 1913 book, *Waves of Sand and Snow*, finding that 'when a patch of sand...is deposited in a fairly exposed position, the surface is smooth and the texture of the sand uniform at first, but almost as soon as rippling commences, the sand on the crests is seen to be coarser than elsewhere' (Cornish, 83). When wind-triggered movement occurs, 'the large grains which accumulate at the crests roll upon the surface, but the finer grains seem to be caught in the air and whisked away' (ibid.). Hugh Miller recounts the strange case of *Jabel Nakous* or 'The Mountain of the Bell...in the neighbourhood of Cabul...about three miles from the shores of the Gulf of Suez.' In a terrain of rock and sand, camel train leaders were sometimes forced to restrain their animals as they responded to 'a strange, inexplicable music. As he leads his camel past in the heat of the day, a sound like the first low tones of an Aeolian harp stirs the hot breezeless air. It swells louder and louder in progressive undulations, till at length the dry baked earth seems to vibrate under foot, and the startled animal snorts and rears, and struggles to break away' (Miller, 59). In less uncanny circumstances, sitting on a beach facing the Atlantic on St. Catherines, a barrier island off the US coast in Georgia, David George Haskell lyrically describes the effect of such movement; sand, wind and varying degrees of moisture, barely perceivable, but in essence, the murmur of a shifting dune's soft pure music:

> Sitting close, I heard this face whisper, a sibilant hesitation, only audible when the seethe of distant wavelets quieted for a few moments. The sounds came from liquefied sand, patches of the slope that suddenly lost their grip and turned, in an instant, to fluid from granular solid. The sand hissed as it raced down the slope in narrow chutes. (Haskell, 62)

Temperature variations and resultant compression and expansion acting as the generators of sound are familiar to us. We have talked of the unsettling effect the timbers of an old house can have on us as they contract and expand at night, just as a fisherman with a full load heading home will know the sound of his vessel, and the subtle signals the weather and the sea transmit through it: the smooth running of the engine, the creaking and groaning of the hull, the slapping of the waves against the waterline. These are reassuring sounds, heard but not attended to; it is change that signals alertness. Similar small sounds exist in dense woodland as the terrain responds to climatic change; ice on rivers and lakes creaks and groans as a thaw sets in, or as expansion occurs. By listening consciously, say with the aid of a microphone and a headset, we may become aware of natural sound in a way that the unsupported ear might miss. Radio could aid us in this, but most radio, as the composer and sound artist Hildegard Westerkamp reminds us 'engages in relentless broadcasting, a unidirectional flow of information and energy, which contradicts the notion of ecology' (Augaitis and Lander, 94). Many sound practitioners prefer to set up listening events with an audience, in order to have greater control over the playback of sound, rather than commit a poor impression of what *they* heard in the field to an audience listening at the mercy of poor frequencies and passive hearing. Nevertheless, it may not be in the audience technology so much as the *intention* behind it that can encourage us to eavesdrop on the small but vital sounds that whisper and murmur around us, and in listening, we may be hearing symptoms of the world as it changes. Westerkamp, while critical of the familiar radio concept of talking *at* us, offers a possible solution through a series of key questions:

> What would happen if we could…make radio before imposing its voice like an alien in a new environment? What if radio was non-intrusive, a source for listeners and for listening? Can radio be such a place of acceptance, a listening presence, a place for listening? Is it possible to create radio that listens, that in turn encourages us to listen to, and hear, ourselves? (ibid.)

Early in 2018, the UK regulator Ofcom announced a new community radio station for Dartmoor National Park. Called Skylark, its intention was to create a soundscape of human and natural sonic phenomena generated by the area and its voices, human and otherwise. The sound of the station was designed to change with the weather, the seasons and the time of day. Recordings made around Dartmoor, uploaded and played out, either pre-recorded or streamed live would be unprocessed and random, a composite of local voices, field recordings, new writing, unaccompanied song and improvised music, entirely recorded within the broadcast range of the station. This random nature of the output, mixed, looped, overlapped and faded in and out would create an ever-changing collage of local life mirrored the inter-species community of the place. The experiment clearly showed that there is a desire in us to not only listen to the often overlooked sounds that surround us, but to interact with them as part, rather than separate from, the world. In our next chapter, we shall return to the sound of self and consider the voice of our own species as its motivating spirit speaks of its being and identity to others—and sometimes listens.

NOTES

1. Boswall, James. *How Birds Learn Their Songs*. British Library. https://www.bl.uk/the-language-of-birds/articles/how-birds-learn-their-songs.
2. Anhäuser, Marcus. *The Silent Scream of the Lima Bean*. Max Planck Institute of Chemical Ecology. https://www.mpg.de/942876/W001_Biology-Medicine_060_065.pdf.
3. Gagliano, Monica, Mancuso, Stefano, and Robert, Daniel. *Towards Understanding Plant Bioacoustics*. http://www.linv.org/images/papers_pdf/1-s2.0-s1360138512000544-main.pdf.

CHAPTER 5

Speak My Name: The Ownership of Syllables

Abstract The sound of a familiar voice can have more emotional impact than a photograph, because it moves through real time as a sonic signature. The ownership of names, words and localised sounds, the use of naming in commemoration and memorial performance events is explored, as is the importance of the reawakening of a phrase or a name that has been lost in history. Thus, the word is an event. On the other hand, words can move beyond meaning into music, and the word as sound can be key to description. The sound of language is central to nationhood and tribal belonging, and it is important to have an understanding of sound qualities in spoken words as the key to personal identity.

Keywords Voice · Words · Identity · Oral history

A Faint Tracing

In 2008, I made a radio documentary about the history of domestic recording, with the producer, Andy Cartwright. As part of the research for this programme, we discovered a privately made cylinder recording of a family sharing Christmas festivities in their home in Salisbury, Wiltshire, in the south-west of England. The year of this rare recording was 1917, the First World War was in its darkest time, and it was clear from the comments on the recording that the father of the family was absent.

© The Author(s) 2019
S. Street, *Sound at the Edge of Perception*, Palgrave Studies in Sound, https://doi.org/10.1007/978-981-13-1613-5_5

We do not know who these people were, and they are all gone now. Nevertheless, as the recording played, they came alive in a touching way, offering as the sound did, a temporal, spontaneous glimpse of lives living a brief moment from a century before. Were we to be able to locate a photograph of this family, it would probably be a formal one; they might be seen posing stiffly, waiting for the camera technology of the time to freeze the moment, and consign it to history. Yet here in their recording, they lived and breathed, laughed and played, sang and joked in a startlingly normal and everyday manner that in one sense removed the strangeness of a far-off time, while on the other, heightened it, making it uncanny. At one point, a young woman is heard to say, directly to the listener: 'Hello. Merry Christmas to you.' Suddenly, coming through the patina of a century-old wax cylinder, the voice of a dead person I shall never know speaks a greeting, seemingly specifically to me. This as I listen now becomes the sonic *punctum* of the experience, and it is written into my memory forever.

Both a photograph and a sound recording play with time, just as our consciousness absorbs visual and aural messages, and both have the capacity to move from the moment as it is lived, into memory. We might also consider that sound itself possesses the facility of imaginative photography, and the most eloquent (in every sense) example of this may be found in our capacity to absorb music. As discussed earlier, a possible explanation for this is linked to the mystery of music itself and its relationship to time, which is constantly moving through a permanent present. Sound—and in this case music—is effectively time made audible, just as the voice may be considered to be air made audible; the experience of it is constantly current. The brain absorbs a newly heard piece of music on both a conscious and a subconscious level, and when this occurs, we are actively absorbed in learning, and processing information that we then store, pending retrieval. As Oliver Sacks has said: 'When we "remember" a melody, it plays in our mind; it becomes newly alive...We recall one note at a time and each note entirely fills our consciousness, yet simultaneously it relates to the whole' (Sacks 2008, 227). As we listen to music, we are hearing the moment, while relating the moment to the immediate previous moment, and moving within a split second into the next moment. It is in other words, very close to the experience of having our attention caught by a striking photographic image. We might take this further into the realm of language and consider the images words evoke pictorially through speech, including meaning but also through timbre, pitch, pace and volume. Within this context, we might view the printed word as a kind of musical notation, a record of thought that manifests itself through the eyes, being absorbed by the brain and then translated into sound, either spoken aloud or imagined internally. It is the partnership between sound and image—the sonic responses produced in us by art and the pictures sound makes—that place us in the world. Once that response has been set up, it has the further capacity to become lodged in memory, thus evoking an image of itself and often, of one's first experience of it. We are ourselves part of a huge audio-visual work, and while Berger may say 'It is seeing which establishes our place in the surrounding world,' (Berger 1972, 7) we might add emphatically, 'and listening.' As a carefully conducted experiment, to open one's front door, or a window onto the world outside in the morning with the eyes closed and the ears and mind alert offers an unexpected field of aural 'vision,' providing stereophonic layers of meaning built of pitch, volume and perspective. To do so actively is to unlock a sense of place and identity that is both startlingly immediate and eerily strange.

It becomes clear that everything is a kind of music. Throughout this journey, we find ourselves reflecting on the idea of the world as soundscape, made up of millions of tiny sonic notes, just as that same world is made up in physical terms of molecules. Phonemes, phrases, sentences and individual letters as sounds build with a rustling leaf, a car tyre on a wet road, a dripping tap or the soft purr of a cat towards a found symphony in which our response to what we hear identifies us as the person we are, and enables us to interact with our environment and our circumstances. As Annie Dillard wrote, 'our life is a faint tracing on the surface of a mystery' (Dillard, 145) and the transience of sound—because it moves through time, as do we, is metaphor, *memento mori* and at the same time part of the genuine fabric of what makes us. Key parts of this web and weft of human personality are the personal sounds that define us, the sounds we own that possess more significance for us than any others, formed as they are of the words, numbers and fragments of song to which we ascribe particular significance.

It is a primal thing; in modern times, we are surrounded by a babble of sound in all forms, and the central element of making sense of this cacophonous world is the ability to be selective. We rely on the multiplicity of sounds—physically heard and perceived within the mind from incoming signals in print, image and impression—to find where we are, how we feel and what our responses should be. Yet, as Walter

Ong has observed, in this process, in which we are complicit, of creating an information technology that both liberates and binds us, we have become distanced from the immediate power of the very sound of language. Or so we might think. 'Early man had no such problem: he felt the word, even when written, as primarily an event in sound.' It is this sound which is 'a natural mystery [that] establishes contact with human existence' (Ong, ix–x). The brain is extremely adept at interpreting sonic shorthand, and in so doing we may create pictures, or sense the instant dart of mnemonic association, taking us to a person, a time or an event. A place or a date may acquire a loaded meaning through an event that lodges in the national or global consciousness; it may be an anniversary linked to a location, as in the French Revolution or a tragedy so profound that it needs no name and place to identify it, such as '9/11'. In those simple numbers are contained an instant reference: they have become the key to a moment in history that changed the world.

It was this moment in time that prompted two of the most extraordinary examples of contemporary American music, composed by Steve Reich and John Adams. These works belong to our story of tiny sound moments here, rather than in our earlier discussion of music in Chapter 3, because in both instances, they involved fragments of everyday speech that gain deep meaning through the circumstances in which they were uttered, and by the provenance of the speakers themselves. Reich's *WTC 9/11*, composed in 2010 is written for a string quartet playing live, over which are placed the sound of two further string quartets, and importantly, pre-recorded voices. In the first movements, these are the voices from archive recordings taken from air traffic controllers, the City of New York Fire Department and the North American Aerospace Defence Command. The second movement, titled '2010,' draws on memories of the events of September, 2001, while the last movement, 'WTC' is based on the acronym of the World Trade Centre, combined with an inspiration from Reich's composer friend, David Long, who in 2003 wrote a piece called *World To Come*. Thus, the power of initials as significant initials is highlighted in a text that could not be shorter, yet could not mean more.

John Adams was commissioned to write *On the Transmigration of Souls* by the New York Philharmonic Orchestra, the Lincoln Centre and an anonymous but prominent New York family, shortly after the 9/11 terror attacks and began writing the piece in late January 2002. It was premiered by the New York Philharmonic, conducted by Lorin Maazel on 19 September 2002. It is composed as a single movement, with a duration of approximately 25 minutes. The work uses taped sound of New York—cars, laughter, a low-level murmur of the city being itself—together with whispered voices, the whimpering of children, the sound of shoes on stone floors, as a barely heard soundscape over which voices intoned the names of victims, spoken like a mantra by family members and friends of those who died. The orchestra plays a tender and sometimes passionate counterpoint to the words, and a chorus sings short phrases such as 'He used to call me every day,' 'My brother' and single words such as 'Missing' and 'Remember.' There is a fine line between the expression of sorrow and mawkishness, but Adams uses restraint and perfect timing in the placement of text, to elevate the work to a truly tragic stature. The orchestral sound is for the most part, virtually static, the effect being to give the stage to the victims themselves, simply by acknowledging the sound of their names. In live performance, it is as though the walls of the auditorium are removed, and the city outside, with its inhabitants, living and dead, enter and possess the space. In his *New York Times* review of the work, published two days after its first performance, and headlined emotively, 'Washed in the Sound of Souls in Transit,' Anthony Tommasini made the point that 'some listeners may find Mr. Adams's material to be insufficiently involving on a purely musical level. But this atypical concert work asks you to put aside typical expectations.'[1] Indeed, the first minutes of the piece contain no music at all in the orchestral sense; emerging out of silence, the sounds of the street filter into our space, and then very quietly, the voices. 'Mr Adams is reticent about calling this work a musical composition. His intent, as he wrote in the program note, was to create a "memory space" where "you can go and be alone with your thoughts and emotions." He wanted to make the concert hall something akin to a great cathedral, where you feel the generations of souls even as you are surrounded by other people' (ibid.).

Above all, it is the familiarity of a name, spoken, that has the most significance, because although it may not be the name of someone we know, we recognise its personal importance, its ownership, to those who spoke its sounds most often. Like our own birthday, and the birthdays of our loved ones, the sound of a person's name is rooted in the everyday. It is the ultimate personal possession, the sound that will turn a head for those who know it. Our own name is our sonic centre, an audio representation of who we are; we can never hear it as a stranger hears it because it is the essence of subjectivity, from which all other ripples of meaning spread outwards. While we may indeed be 'a faint tracing on the surface of a mystery,' the marks we leave, or which are left in our memory, whether they be written or spoken, prove our existence.

At the Grodzka Gate

In the opening chapter of this book, we explored the significance of the insignificant, when nothing else survives. The tiny bell of the Lublin poet Jozef Czechowizc rings in the mind as a link with the man, and the Poland that he heard happening outside his window before the Second World War. That same bell rings with another memory of the same war. In Lublin, there is an ancient portal that once was the transmission point between the Christian and Jewish districts of the city. It is called The Grodzka Gate; on one side, there is a busy pedestrianised street full of bars and restaurants, while beyond it, on the way to the great castle on its hill, there is what local people call 'the great space.' This was where, prior to 1942, the Jewish town existed, with its population of approximately 43,000 citizens. Today, there is no Jewish district, indeed there is nothing to show it ever existed. From 1942, the occupying Nazi forces obliterated all physical trace of it.

In 1992, an avant-garde theatre company, the 'No Name Theatre Company' took over the Grodzka Gate, with the intention of staging studio style drama within its spaces. When the company discovered the story of the lost Jewish town, its purpose changed. Led by its director, Tomasz Pietrasiewicz, the role of 'Teatr NN' became one of witness, of chronicling and recording every lost street, every missing person, every name and memory that could be pulled back from the dark time. Today, the visitor to the Grodzka Gate becomes part of an immersive experience in which 'the great space' becomes peopled once again with living names. As part of this, the visitor is encouraged to select a name from the vast catalogue of humanity that is now here recorded, and, as it were, to adopt that name, and speak it aloud, because by speaking a name, by vocalising that person's most intimate identification, their presence moves through time once again. Likewise, we write the name by hand and carry it away with us, because writing by hand is a physical, tactile act. It is human. You can *hear* a signature, a hand-written word.

This is not merely metaphor; it is *literally* true. As Rober Racine has said, 'handwriting possesses a rhythm, a cadence, a sound of its own.

When writing, each individual creates a unique music. Writing on paper, slate or heavy cardboard with a felt pen, graphite pencil, chalk or charcoal stick creates new tones that resonate with acoustical precision' (Racine in Augaitis and Lander, 141). In 1992, Rober Racine conducted experiments at the Banff Centre's Electroacoustic Recording Studio in which twenty-two individuals were invited to record the sound of them writing their signatures. To listen is to 'witness the acoustic expansion of a secret, intimate and barely audible world. The writing lives and breathes...' (ibid.). When the recordings were broadcast, they were changed by the very act of transmission into something even more poignant, resembling 'a faint whispering, a gentle presence that reconnects with the original intimacy of the handwriting' (ibid., 142). To write by hand, particularly to write a name, which is a badge, a brand, a sound that is or has been familiar to its owner in a way it cannot be to anyone else, is to enter a partnership with sound, and in this instance, with radio itself, a solitary and secretive thing: 'It is first and foremost the incarnation of a privileged listening experience, a sort of confidant' (ibid.).

Such an experience can exist in reverse; as we have seen, when we read, we engage sonically with images implied by a writer through a series of codes. Put simply, because we see words, we have the capacity to imagine, and through that capacity, the potential to *hear*. Engaging with a text from a long-dead writer, we participate in a form of imaginative resurrection, as the text connects us to a human life that made it in real time, as Mark Doty has suggested:

> The form, spoken, breathes something of that life out into the world again. It restores a human presence; hidden in the lines...is the writer's breath, are the turns of thought and of phrase, the habits of saying, which make those words unmistakable. And so the result is a permanent intimacy; we are brought into relation with the perceptual character, the speaking voice, of someone we probably never knew, someone no one can know now, except in this way. (Doty, 50)

At the Grodzka Gate, and at other places of human witness, by speaking a name aloud, we create the radio of memory, wireless transmitters as we are. Theatr NN is as much about oral history as anything else. The Gate now houses a huge oral history archive, and as Pietrasiewicz writes, by 'using it, we want to emphasise that the registration of memories becomes a specific moral commitment towards the witnesses of history – we do it to ensure that the bequeathed stories will be safe and will remain alive' (Pietrasiewicz, 412). It begins with the sound of a long silent name spoken again.

WORD AS EVENT

The phrase is borrowed from Walter Ong (Ong, 111–37). In his book, *The Presence of the Word*, he discusses the idea of auditory synthesis, and the direct link between sound as a temporal phenomenon and our use of it as an interactive tool with which we develop our own identity is poignantly linked to our own transience. Ong makes an important point when he says this, (the italics are his):

> *Sound is more real or existential than other sense objects, despite the fact that it is also more evanescent.* Sound itself is related to present actuality rather than to past or future. It must emanate from a source here and now discernibly active, with the result that involvement with sound is involvement with the present, with here-and-now existence and activity. Sound signals the present use of power, since sound must be in active production in order to exist at all. (Ong, 111–2)

In order to speak we are consciously breaking a silence, making a commitment to a discourse, entering into a debate and fundamentally establishing a presence. Once spoken, the word cannot be unspoken, and the event of a word placed into time is thus a profound establishment of our existence within the flowing river of universal history. Faint, brief, faltering it may be, but once uttered, whatever its meaning, its sound is undeniable. The distinguished voice coach, Patsy Rodenburg, has pointed to the bodily act of speaking: 'Any voice, in order to make and project the appropriate sounds, requires the full use of a set of interrelated physical acts. The very sound we make is a result of physical exertion' (Rodenburg, 20). The materiality of this act of creation is the beginning of the mystery, but the greater wonder is in the transfer of thought through the physical and back into memory. Key is the *way* the sound forms and interacts with its acoustic surroundings, and with the listening presence. Just as it is often the smallest details, the fragments of personality and experience that matter most within the biggest events, so memory may amplify (as opposed to exaggerate) the significance of tiny sounds. On the printed page, we may use italics to drive home a point we wish to emphasise; in speech we may be more subtle, and a shouting voice has far less emotional longevity than a whisper. We underestimate the power of sound nuance all the time.

In 2016, the radio producer Eleanor McDowall created an initiative called *Radio Atlas*, a partial solution to the problem of interpreting international sound documentaries and features in languages other than one's own, an English-language home for subtitled audio from around the world, opening cultural possibilities and appreciation of creative audio: inventive documentaries, dramas and works of sound art from programme-makers across geographical and linguistic divides. Like all great revolutionary ideas, it is remarkably simple in concept and received a Special Commendation at the Prix Europa in 2016. Yet as Emma Smith explained with regard to her *Euphonia* project earlier in this book, there are moments in sound that require no translation, and we shall return to this in our final chapter:

> Those moments of silence – the musicality of delivery, the deep breath, the held moment – are the reason why I think translation struggles when it's imbedded in the sound as opposed to using subtitles. Doing this over and over again with different documentaries makes me realise how those are the points that carry the meaning, the deeper understanding, not just the facts of what's being said but what they mean to the speaker.[2]

It is fundamentally to do with the passage of sound through time, its medium of air as opposed to light and the inherent longing we all possess to tell our story, to make a connection with another spirit—in short, to communicate. 'In radio, speakers often deal in reflection, past tense recollections, but I feel those spaces between words somehow contain the present tense feeling' (ibid.).

SYLLABLES AS SOUND

Both the composers Mussorgsky and Rimsky-Korsakov observed that music and speech are closely linked. It is interesting to observe that both were Russian, and one can speculate as to the relative euphony of certain languages more than others. Voice recognition is inbuilt into us, and the empathy with the sound of a person is a major part of the attraction between individuals. We can long to see a beloved face, but at least equally, to hear a voice is to sense a life. Just as a phrase from a song instantly connects with a part of our brain that identifies with it in a profound and intimate way, entering the fabric of our own personality, so the sound of a voice, its timbre, pitch, modulation and tone can put us at our ease or disarm us, attract or repel our responses. Within minutes of their first meeting, Juliet declares to Romeo, 'My ears have not yet drunk a hundred words/Of that tongue's utterance, yet I know the sound' (Shakespeare: *Romeo and Juliet* Act 2 Scene 2). Certain fragments of speech contain a modulated sound that is literally music, and the repetition of a word or a phrase, or playing it on a looped recording may produce the disorientating phenomenon of loosing meaning while creating a tune. Indeed, the work of the Canadian composer Priscilla McLean, which we shall explore shortly, exemplifies this. To demonstrate the idea of sound patterns as a kind of song, it is only necessary to choose a short phrase of five or six words and speak it out loud continuously. Diana Deutsch, professor of psychology at the University of San Diego, California, has experimented with this and other ideas, using recordings. While playing back a recording of a talk she had written, she noticed that, while discussing words, she had commented that 'they sometimes behave so strangely...' Listening repeatedly to the recording, the phrase began to lose its meaning, while at the same time, rhythm, pitch and timbre seemed to take over to create a short melody that she was then able to notate and play on her piano.

Steve Reich had taken this idea into the concert hall with his piece, *My Name Is*, created as long ago as 1967, a work for which there is only a verbal score rather than notation in the traditional sense. It is, in a sense, an ultimate expression of participation; as the audience came into the hall, a member of Reich's team stood at the door with a small recorder and invited concertgoers to speak their Christian name, introduced by 'My name is...' Thus: 'My name is X,' 'my name is Y' and so on. Then, as the concert progresses, Reich would take the tape and choose half a dozen or so names that he felt were the most euphonious and melodically interesting, making three identical tape loops from the names. After the interval of the concert, the piece would be performed, with the recordings played back in a phased sequence, starting with a particular name thus: X-XX-XXX, which would then fade, to be replaced by Y-YY-YYY, etc., before next mixing them in sequence, creating a pattern of identity that reverberated around the hall:

Psychologically, the result was interesting because they usually said their name in a very off-hand way because it was a funny thing to do when you walked into the hall…Hearing your name in that way tends to get to people – its sort of like doing a sketch of people at the door… (Reich, 29–30)

An even more radical example of word repetition has been explored by the composer and sound artist Alvin Lucier, in his piece, I Am Sitting in a Room, for voice and tape, first recorded at the electronic music studio, Brandeis University in 1969. It was subsequently performed in various other recordings, culminating in 1980, when Lucier made a version in his own home. The piece begins with a straightforward reading of a text, written by Lucier, in which he states his situation, sitting in a room, recording the sound of his voice as he tells the listener that the intention of the exercise is to then replay the sound repeatedly until the frequencies of the room itself assert themselves increasingly, leaving the human aspect of the speech, apart from a sense of its rhythm, no longer perceivable. The reading is straightforward, undramatised and unemotional, it is plain words, stating fact and intention.

What we hear is just this spoken paragraph within the acoustic of the room. The recording then goes through its cycle of repetition 32 times, being added to by each new version cumulatively as it does so. The effect of this is to draw out and resonate each syllable, to the point that, by the end, it becomes impossible to distinguish where one word ends and another begins, the text being completely unintelligible. What was once a familiar word has become an eerie three-note motif, and the simple mundane statement has become purely tonal: music. Over a period of approximately 40 minutes, meaning has gradually shifted from language to harmony. Yet because we know the original statement, it is possible for the attentive listener to retain the essence of what was said, morphed though it is by acoustics, memory, feedback and the presence of the room, and this creates a strange feeling within the mind that the personality of the speaker has somehow over the period of the experiment become one with the character of the space itself.

The syllables, the notation of language can send signals to the mind to enhance or even replace less eloquent sounds as we observed in chapter three. In this idea of pure sound as meaning may lie deep-seated prejudices and inherited fears, as a youthful Patrick Leigh-Fermor experienced on arriving in Germany on his epic walk as an eighteen year old from

Rotterdam to Istanbul in 1933. In his account of that moment in his remarkable book, *A Time of Gifts*, published more than 40 years later, he graphically analyses his response to the euphonic structure of a single word:

> Germany!…I could hardly believe I was there.
> For someone born in the second year of World War 1, those three syllables were heavily charged. Even as I trudged across it, early subconscious notions, when one first confused Germans with germs and knew that both were bad, still sent up fumes, moreover, which the ensuing years had expanded into clouds as dark and baleful as the Ruhr smoke along the horizon and still potent enough to un-loose over the landscape a mood of – what? Something too evasive to be captured and broken down in a hurry. (Leigh-Fermor, 35)

It is salutary to consider how deep such associations go: beyond the word itself, deep into the psyche. We are surrounded everyday by signals coming to us from media and popular song that convey subliminal messages and produce subconscious responses in us as we go about our business or as we drive the car. 'America first,' 'Brexit means Brexit,' 'This lady's not for turning,' 'Ich bin ein Berliner.' Politicians—or at least their speech-writers—know how to plant a seed… Language as sound is the shorthand of thought. Listening to a language of which we have no knowledge is the ultimate exploration of speech as music; relieved of the necessity to understand, we listen with an innocent ear to the rise and flow of emotion, joy, humour, pain and suffering, all communicable in sound beyond nationhood. We hear with our ears, but we listen with our minds, and our antennae are finely tuned to respond to the sounds of what we need—or want—to hear. Diana Deutsch has observed this, referring to the phenomenon as 'phantom words':

> People often report hearing words that are related to what is on their minds. If they are on a diet, they may hear words that are related to food; if they have had a stressful day they may hear words that are related to stress; and so on. In fact, so strong is the influence of meaning on what is perceived, that people sometimes hear voices speaking in strange or unfamiliar accents, so as to create for themselves words and phrases that are particularly significant to them. (Deutsch, 2)[3]

We come back time and again to the writings of John Clare and his quest for precision of sound in language as expression. Clare's precision of thought and frustration at the lexicon available to him sometimes led him to make up or appropriate words if their sound matched his purposes. In John Clare's poem, 'The Flight of Birds', words we may not recognise in this context are given new life and meaning. For example:

The Pigeon suthers by on rapid wing.

The word 'suthers' as a verb is archaic, meaning either the sound of the wind sighing or moaning, or—as Clare uses it here—as the whirring of a bird's wing. Yet used to identify a pigeon's flight, it does more: it gives us the essence of 'pigeon' in a sound, not the call of the bird, but its inner character. Because I hear it, I see it. Earlier in the same poem, he is equally precise:

The crow goes flopping on from wood to wood...

No one has seen or said it better than this. Then, in the next line, there is this:

The wild duck wherries to the distant flood.

The Shorter Oxford English Dictionary gives me: 'Wherry (1443, of unknown origin): a light rowing boat used chiefly on rivers to carry passengers and goods; a large boat of the barge kind (1589).' The way Clare uses it however gives us the sound of a duck as it takes off or lands on water to perfection by turning it from a noun into a verb. Living on the edge of the Fen, he would have known the wherry as a boat plying some of the waterways of East Anglia, particularly in Norfolk. Which leads me to ask if the word's original definition was based on the sound it made as it scudded through the water? 'Wherry' is full of the breezy light and open skies through which Clare's birds flew, and the chill waters upon which the ducks landed; onomatopoeia translated into the object, waiting for Clare to turn it back on itself again. It is at least a possibility. Be that as it may, in such poems and word usage, he teaches us to examine language for its sound as much as for its meaning. To use the word as a noun, 'like a wherry' would have been good: good enough for a lesser poet. Yet as a verb in the hands of John Clare, it transforms meaning through its sound.

The acquisition of a sound bank of associative audio signals linked to meaning and sense is a growing and evolving thing. Augustine wrote that 'if someone hears a sign that is unknown to him, for example the sound of a word of whose meaning he is ignorant, he desires to know it; that is, he desires to know what idea this sound evokes...but it is necessary that he already know that it is a sign, that the word is not an empty sound, but a sound that signifies something' (Augustine, *De Trinitate*, quoted Cavarero, 33). It is this that links the abstract to meaning. The appropriation and ownership of sound are a kind of profound personal democracy.

In his book, *Landmarks*, Robert Macfarlane has created both a guide to writings about the landscapes of the UK, and a dictionary of associated words, many unfamiliar outside of their locality. It is remarkable to realise how varied syllabic expression can be even within the compass of a small island nation such a Britain. Yet the variety and richness can be astonishing, even within one language, spoken regionally. 'Gulsh,' for example, a Northamptonshire word that Clare would have known, perfectly describes its definition, which is 'to tear up with violence, as a stream when swollen with floods' (Macfarlane, 121). Likewise, what could better describe the act of huddling to shelter together miserably from the cold than the Herefordshire word, 'Crool'? (ibid., 225). Or 'bruckle' to convey an 'easily crumbled, stony soil' (ibid., 286). This acquisition of language is an ongoing part of the societal evolution; no language is above taking words or sounds from another.

Yet as language evolves and usage alters, words move out of the realm of the everyday and many eventually disappear. Thus, the sounds of language shift and vanish, as the lexicon of human speech changes. Words such as those cited here may still be in regional use, but many others having fallen out of common parlance altogether, and have become by very virtue of this fact, sounds at the edge of perception—and beyond even that. There is a delight in finding a 'new' word from the distant past that preserves an exact meaning for which no other will do. Followers of the fourteenth-century philosopher-theologian, John Duns, known as Duns Scotus, coined a word that encapsulated his thinking: 'haecceity.' It is a good word for our purpose, insofar as it was used to denote the discrete qualities, properties or characteristics of a thing that make it the *particular* thing it is. It comes from the Latin, 'Haecceitas,' for which the nearest English translation would be 'thisness.' It is 'thisness'—'haecceity'—that describes the impact on us of seminal sound moments. Asked to describe the unique effect on the senses of a sound or a phrase of music, we might fall back on the answer supposedly given to a journalist who wanted a trite explanation of jazz: 'If you gotta ask, you ain't never gonna know!'[4] A word may vanish, and with it a sound, and a precision of meaning or description too. There is a joy and a responsibility in the idea of bringing such verbal music back from the edge of oblivion.

When we consider such relative sound values of international language and examine the phonological and phonetic relationships within a language or between different languages, we begin to understand our human connections. What links us is the air that we breathe because the voice is formed out of the body's interaction with the atmosphere that gives us life. So in the real and metaphorical mystery of the relationship between the physical (the larynx, the articulating organs, the oral and nasal passages and sinus cavities) and the invisibility of the breath that powers them lie, the fundamentals of vocalised communication. To speak a word is to do much more than make a sound; it is to immediately articulate history, culture, nationhood and regional character, and to declare an identity. The American composer, Priscilla McLean, in her work, *Xakaalawe/Flowing*, created in 2004 at her New York studio, fused a number of forces, including conventional instruments—flute, saxophone and based clarinet—with natural sounds, and words from Native American nations, namely the ancient Crow and Arapaho tribes. The inspiration for the piece came during a residency at the Rocky Mountain National Park in Colorado, in which she focussed on the sounds of elk, waterfalls and other distinct sounds found in that particular place. It was, from this standpoint in the context of her work, a relatively short and logical step for MacLean to work on language as a part of the environment, and in particular—poignantly—the expression of common connections of displaced peoples with their environment. She chose specific words to signify these connections between the human and the natural worlds (links that within Native American culture itself are self-evident anyway). Thus from the Crow language, *xakaalawe* (flowing), *aashe* (river), *awaxaawe* (mountain) and *ammiiwishe* (rocky), while from Arapaho she used *choowoowo'oo'* (flowing water), *niicii* (river), *necceensisie'* (waterfall) and *hoho'nookeeno'* (rocks). The mere appearance of these words on the page—even before one applies any phonological considerations—demonstrates the tribal differences within relatively adjacent geographical peoples. There seems—to the exterior observer—to be no perceivable link for example between the crow word for 'flowing' and the Arapaho word for 'flowing water.' Yet these are written interpretations of sounds, and as such their origins lie deep within phonic history and spoken identity. The most significant connections between the cultures of tribes exist in a common understanding of unity with nature.

English speakers may find the sonic variations in certain languages—and their representations in their respective alphabets—to be at first bewildering and perhaps somewhat intimidating. Nevertheless it is instructive to consider how differing phonetic criteria govern meaning in ways that make spoken English, for all its richness and variety, seem almost crude. Chinese, for example, is a tone language, a language in which changes of pitch of the voice are used to denote differences in word meaning. 'We can think of tone as a third type of speech element in addition to consonants and vowels. In general, each Chinese syllable bears a tone' (Yen-Hwei Lin, 3–4). We should remind ourselves again that language is essentially music, and each morpheme (the smallest meaningful unit in a language) may be thought of as a single note in that music. 'A Chinese word like *xuéxiào* (school) has two syllables and two tones, a high rising tone on the first syllable, *xué* and a high falling tone on the second syllable, *xiào*' (ibid., 4). The extreme subtlety of the language may be seen in the differences of meaning with the simplest of single syllable words, 'ma.' In this case, varied by pitch patterns—high/ high rising/low falling-rising and high falling—and a range of specific tones, using phonetic script, Má means variously 'hemp,' *mǎ*, 'horse' and Mà, 'to scold.' These simple and basic examples are enough to point to a realm of linguistic possibilities for subtle expression. Yet it goes beyond vocabulary and spreads into religion, philosophy and ancient cultural traditions, so that to properly interpret the meaning within a Chinese poem is perhaps beyond translation, just as to uncover the world within a tiny haiku by the Japanese poet, Basho may only approximate to the full shading of the poetry and what lies between the sounds.

Always we find ourselves returning to the voice of the poet, because it is here that a distillation of language as sound becomes an essence that is powered by the minutiae of expression, in turn crucially lending itself to identity, as we remembered Eliot expressing it in an earlier chapter. By 'drilling down' into the component parts of words and phrases of one language, relative to another, as in this conversation with the Chinese-/English-language poet, Jennifer Wong we uncover some key comparisons:

J.W.:   The tone system of my language is very different from that in English of course, although I'd say the prominence of rhyme in Chinese poetry (classical) has helped me to be aware of sound and to make it serve sense. I suppose it's a bit easier to find rhyming words in my native language because I know it well and that it has 4 tones (Mandarin)/9 tones (Cantonese). Then there are the "tone categories." Also, lots of Chinese words can sound a bit similar but not the same, so I guess it generates the sound of half-rhymes more. That tends to be quite satisfying I think.

S.S.:   So is it a question of 'less is more'?

J.W.:   In Chinese poetry there is the possibility/advantage in being able to say a lot more with a few words, hence the lines in a poem can be much shorter yet still as powerful. Whenever I can, I tend to apply this principle of saying more with less in English. I suppose this training is actually good for poetry-making as it makes me very aware and keen to pack more meaning into each word. On the other hand I do find it more challenging when writing English poems.

S.S.:   Can you say something about the role of syllables?

J.W.:   As most of the time English poetry writing is concerned with stresses rather than syllabic counts, this inspires a different way of writing and understanding poetry. In Chinese, we tend to count syllables as each character is a syllable unit.

S.S.:   Coming back to sound, and linking it to meaning, what would you say is the relationship between the two languages in dealing with the elliptical, that is to say, the ambiguous versus the specific?

J.W.:   I enjoy the specificity of English words as each word tends to have a more specific meaning, so when it comes to writing poetry, precision becomes quite easy to practise. On the other hand, each Chinese word can easily mean a few things depending on context, which is also a different sort of strength when it is used in creative writing, because I can be meaning several things at the same time by removing a very obvious context for it.

S.S.:   Does this extend into grammar and syntax?

J.W.:   Chinese grammar is less concerned with the consistent correlation of tenses. As such, it is quite liberating for creative writing, hence I do find it more restrictive when it comes to writing poetry in English because things have to "make more sense" and, while grammar or syntax can be changed, there is still an overall need to be grammatically "correct."[5]

Turning to another great poetic language, such as Arabic, we find that while some sounds differ from English, while others do not possess an English equivalent at all. Vowels may be long or short, while a long vowel is held twice as long as a short vowel, and these vowels may be sounded differently according to the consonants that surround them. It is salutary to understand the variations in global language, if only to view our own word usage objectively, and to realise how confusing and illogical are some of the rules that we take for granted, having absorbed them from birth. We have mentioned time as an element in language, and it is relevant here because all languages evolve. To examine for example, the way language is used by some of the ancient writers and philosophers of Persia/Iran, such as the writings of the Shiraz poets, Hafez, Jahan Malek Khatun and Obayd-e Kakani, is to encounter modes that have changed and evolved just as English has moved from the vocabulary, style and dialect spoken and written by Chaucer. There are jokes and puns written into the plays of Shakespeare that are often lost on a modern audience, because they referenced a vernacular that has changed since his day. Above all, we return time and again to the *sound* itself: The twelfth-century Persian poet Shatranji wrote that 'The beauty of a verse is in its rhyme' (quoted in Davis, lxvii). We all make our own music, whether or not we understand its history. Just as 'modern poetry in English has many forms (the sonnet, the ballad, the heroic couplet and so on), but only two metres that are at all common (iambic and its derivative iambic/anapestic, and trochaic); Persian poetry uses in all twenty-three different meters...but only two forms – monorhyme and the couplet' (Davis, lxvii). While the intricacies of variants within international poetic form may be not the main concern of this book, it is relevant to use poetry as an exemplar for identity since its nuanced consideration of sound as meaning becomes so often the voice of national and tribal emotion and expression. To 'hear' one of the great gardens of Shiraz, for example, one has only to read Hafez; even in translation, the relishing of beauty in the natural world shaped by man within the broader context of arid lands becomes both a powerful form of identity and a metaphor.

All this is to suggest that 'naming' is a part of our identity far beyond the simple badge applied to us at birth, and which we carry with us through life as a representation of 'self.' The sound of our words and the way we say them—fundamentally our own personal song—reflect who we are, where we are and where we were planted. Music, as we have seen, is as much a form of articulating identity as the spoken word, so we may take 'word' in this respect as sound, in the sense that it is an utterance uniquely our own. It becomes fundamentally about breaking the silence to establish our right to exist because 'confronting silence by uttering a sound is nothing but verifying one's own existence. It is only by singling out one's self from the cavern of silence that can really be called "singing"' (Takemitsu, 17). The Welsh language metrical sound form of Cynghanedd, sometimes known as consonantal rhyme, is a patterning of consonants, rhyme and stress; the word actually translates as 'harmony,' something to be heard rather than written. For the human species, the smallest sound can link person to person, initiating life-changing processes, as the poet Clare Shaw, who also works in mental health training, has written: 'Blessings, songs and curses; rhymes, spells and prayers – combinations of words and sounds which are profoundly beautiful for their own sake – and which can in themselves, bring healing or illness, love, wealth or death. Which can make or break a person, a history, a people' (Shaw in Ivory and Szirtes, 187–8). Both as a communication of thought and of pure feeling linked to memory and inner sense, the human voice is an instrument of infinite subtlety and nuance, creating an instant bridge of interaction that can touch the spirit in a most profound way. No wonder the Little Mermaid in Hans Christian Andersen's story says 'if you take my voice, what shall I have left?' (Andersen, 153).

NOTES

1. Tommasini, Anthony, in *The New York Times*, 21 September, 2002. http://www.nytimes.com/2002/09/21/arts/philharmonic-review-washed-in-the-sound-of-souls-in-transit.html.
2. McDowall, Eleanor. Correspondence with the author.
3. Deutsch, Diana, *Phantom Words and Other Curiosities*. CD, San Diego, La Jolla: Philomel Records, 2003.
4. The origins of this statement, like many others belonging to the oral history of jazz, is hard to identify. It has been variously attributed to Louis Armstrong and Fats Waller.
5. Interview with Jennifer Wong, 3 March 2018. Reproduced with permission.

CHAPTER 6

First and Last Sounds: Messages Beyond Language

Abstract We remind ourselves of the variations of global voices, and we come to the smallest sounds defining a place, a person or a time. We revisit the moment of the first transatlantic wireless signal, to underline how a single morpheme may be the key to unlock a revolution. Radio producers share their experiences of catching an essence from the passing air. There is sound caught inadvertently that may be more significant than the object of a recording, history captured as it were, in the corner of the ear. Sound may take us back to childhood, and if it is true that it is the first sense to awaken at birth, and the last to desert us when we die, then it is indeed the gatekeeper of life.

Keywords Transmission · Morse code · Global sound · Location sound · Mortality

Mnemosyne: The Sound Under the Sound

Early writings on broadcasting shared an awed concept that there was within the 'ether' something that brought the technical and the metaphysical together. Cecil Lewis, the British Broadcasting Company's first Organiser of Programmes, who had come from the world of pioneering aviation to broadcasting, saw within two years of the birth of the BBC, the potential for the medium to expand literally without limitation:

© The Author(s) 2019
S. Street, *Sound at the Edge of Perception*, Palgrave Studies in Sound,
https://doi.org/10.1007/978-981-13-1613-5_6

The human voice has annihilated space. It becomes endowed with the infinite range of light, for the concerts of last year are still spreading out and on beyond the range of the visible stars. Wireless waves, like light waves, travel at an incredible speed and have greater penetration. Thus, when we speak, it is not to the listener, or even to the world, but to the universe. (Lewis, 144–5)

Sound and light are capable of travelling great distances, but the human imagination has the potential to be limitless. In his book of essays and short stories, *The Moment Under the Moment*, Russell Hoban wrote of his custom of listening to music broadcasts from All India Radio while relaxing in his London apartment:

Sometimes reception is beautifully clear, and the chromatic splendours of the classical Karnatak style build palaces of sound all round me in my Fulham workroom. Swaying painted elephants and iridescent peacocks., chanting priests, multitudes of worshippers, solitary mystics and astronomers, saffron-veiled beauties and dancers with ankle bells glisten in the misty drizzle of the London night outside my window, all India vivid with my ignorance. Great wild eastern dawns and screaming birds rise where the red and green lights of the District Line wink to the passing of the golden windows rumbling townwards, rumbling homewards. Distant passengers, perhaps seen every day, perhaps never to be seen again, pass in the passing windows among painted elephants and the clash of ankle bells, the marble and the filigrees. (Hoban, 217)

In those pre-Internet days, listening to distant cultures on radio was subject to all sorts of interference and signal variance, as Hoban implies; short wave was the main source of contact, and its closely packed stations would fade in and out of earshot at the whim of atmospherics. Today, we can have the sonic world in our pocket, experience curious shifts of time and place as we listen, say to sounds invoking Hawaiian seashores on a headset, while inhabiting the visual world of a rain-soaked Monday commuter run to work. The technology may obscure the wonder of it, but the idea that coming through air are human messages compressing space and sharing time in this way is nonetheless something of which Cecil Lewis and his colleagues could not have dreamed.

We have discussed in these pages that the most poignant and obviously personal aspect of speech—indeed of all vocalised sound—is that it forms itself from the very air we breathe, that it is formed of the breath of life, and that when we speak, we use language to translate the world and respond to it. Language is interaction, but interaction is more than language in terms of aurality. Speaking and understanding spoken language are two parts—call and response—of the process by which we map meaning onto linguistic forms and vice versa.

The first poems were spoken before they were written, and the sound of song is the ultimate expression of humanity. It is far beyond the literal meaning as expressed within the vocabulary itself; how we compile the chain of sounds that become words, in other words, HOW we speak as much as WHAT we speak, conveys our selves and our state of mind. First sounds, even before words, show the world who we are; a hesitation on the edge of expression, a stammer, a pause may speak as eloquently—or more so—than the word or phrase that follows it. Society judges us constantly on first verbal impressions. Dialect, accent, an unfamiliar regional voice or the sound of a person speaking in a language that is not their native tongue will be enough to turn heads in certain circumstances. Equally volume, tone and pitch, whether too loud, too quiet, too high or too low, nasal or harsh, are all factors by which we make judgements. Our antennae are highly tuned to react, however subtly, to the signals that surround us.

Language may require translation, depending on our linguistic skills, in order to convey ideas. Yet emotion requires no translation. When the ninth-century Buddhist monk poet Kūkai suggested that the first utterance was 'Ah!' as an expression of wonder (Hirschfield, 81) he was declaring an idea of great beauty and poignancy that in a very real sense touches on our basic humanity. Kūkai was a calligrapher, and his craft taught him the importance of the shapes of the world, the smallest touches of the pen or the brush on the parchment. So it is with the voice. This idea calls to mind the final words of 'God's Grandeur,' a sonnet by the nineteenth-century English Jesuit poet, Gerard Manley Hopkins, punctuated as it is with sounds of ecstasy:

> And though the last lights off the black West went
>   Oh, morning, at the brown brink eastward, springs –
> Because the Holy Ghost over the bent
>   World broods with warm breast and with ah! bright wings.
> (Hopkins, 139)

The first and last sound we utter—a baby's first cry or the dying sigh at the end of a life—are the most important sounds we will ever utter, an announcement of a new presence and a final farewell to the world. Tom Lanoye wrote a moving novel based on the life of his mother, a vibrant, garrulous, intelligent woman, who loved language and performed often as an actress. Lanoye called his book, *Speechless*, and its theme is that of the predicament a person enters when the power of speech deserts them. In Lanoye's case, his mother, Josée was a flamboyant, domineering, somewhat controlling woman, but she suffered a severe stroke through which she lost her ability to speak. With the loss of her voice, so dear to her, she slowly and with a sad inevitability, declined. The book is a curious hybrid between documentary and fiction, and it ends at Josée's deathbed, where Lanoye notes that 'she is not breathing but rattling. The doctors have assured me that such rattling is normal, not painful and even inevitable' (Lanoye, 344). These are last sounds at the end of a lifetime of vocal dexterity and expression; the last sounds of a human being before silence. At her bedside, her son vows to himself that he 'will combat the silence with my voice, will try to out-argue silence with my word' (ibid., 345). It is as Caliban says in Shakespeare's *The Tempest*: 'This isle is full of noises' (Act 3, scene 2), but the most crucial, the most important of all, are the small sounds with meaning far in excess of their brevity or volume. The microcosm of utterance is like the workings of a ship; to use an analogy expressed by John Durham Peters in his book, *The Marvellous Clouds*, 'means that are apparently small – compass, log, and point – deserve a place in our thinking about that which is great…The ship is more than its moorings, but without them, it drifts or crashes. Moorings are the means that hold the ship where it should be. Being needs such holders' (Durham Peters, 114).

THE HISS OF *YES*

On 12 December 1901, on Signal Hill, St John's, Newfoundland, Guglielmo Marconi received the first transatlantic radio signal, sent from his transmitting station at Poldhu in Cornwall, some 2000 miles away. The process of proving wireless telegraphy in its infancy was a long and tortuous one; cable communication across oceans was already in place and had been achieved at great cost and human effort. The idea that signals could be sent without cable or wire was a concept that had occupied the thinking of many scientists before Marconi, and the idea had moved beyond theory. Nevertheless, as with all pioneers, the surmounting of each obstacle is in doubt until it has been achieved. Cable transmissions could not help shipping; ship-to-shore communication required technology that was reliable, robust and airborne. The answer, wireless, was by its very invisibility for many at the time, a medium approaching the supernatural in its strangeness. Oliver Lodge, the British physicist and a professor at Liverpool University, had established the importance of tuning to wireless communications and even patented a circuit for this purpose as early as 1897. Lodge believed that such was the power of this uncanny force; the possibilities might include communication with the afterlife.

In December 1901, however, Marconi was concerned only with bridging the Atlantic, which to many of his contemporaries seemed almost as inconceivable as speaking with the dead. Yet the groundwork—if one may use the word in terms of an airborne technology—had been done. In various parts of the UK, his team had successfully conducted experiments, including crucially, transmitting across water. The importance of the Poldhu/Newfoundland project was that it sought to solve what many felt was the insoluble problem of wireless waves and the curvature of the earth. It was still thought by some that a transmission sent over long distances would simply travel out into space. What 1901 proved to the world was that there were layers within the Earth's atmosphere that would prevent this, and that a signal would bounce between Earth and sky onwards towards its intended destination.

The decision was to transmit the single sound of one letter—S. The reason for this was an entirely practical one; it is important to remember that what Marconi and his team were transmitting was not the phonetic sounds of speech, but Morse Code. The first wireless signals were therefore not telephony but telegraphy, and the choice had fallen upon the letter S, because it was considered that anything else would have consisted of a series of dots and dashes that, through the white noise of electrical static and interference, could have run into one another. For the reason, the easiest letter to distinguish was S, the Morse code for which is simply three dots.

Three tiny dots travelling 2000 miles through all that space, breaking the silence. S for silence, S for sound. As an Italian, whether or not Guglielmo Marconi was aware of the happy coincidence within the English language that the word 'listen' is an anagram of the word 'silent' has not been recorded. There is a poetry to the idea of sound, but that

December, the issues were physical, scientific and intensely practical: those of maintaining fragile transmitting equipment in working order, through intense storms in two of the most exposed places on either side of the Atlantic: Cornwall and Newfoundland. The place decided upon as the receiving station was Signal Hill high above the narrow entrance to St. John's Harbour, with its imposing landmark of the Cabot Tower. The name of the hill was coincidental; it had no significance for Marconi's experiment other than the altitude, which made it propitious for his purposes. The name in fact originated from the pre-wireless requirement of signalling to incoming shipping with the use of flags, and for transmitting, via the same method, the identity of these ships inwards to harbour officials.

A key witness that day was George Kemp. Kemp had served as an electrician and instructor with the Royal Navy, before working for the post office. From July 1896, he devoted himself entirely to wireless telegraphy, and in 1897 he moved from the post office to the fledgling Marconi Company, where he worked as first assistant for the next thirty-six years. Kemp was at Marconi's side for all his most memorable achievements, and he was there as the December gales fought with the fragile kite to which the receiving aerial was tethered. At 12.30 p.m. on 12 December 1901, the two men were listening intently. Then came the three dots. Marconi waited and it came again. His own account of the moment underlines its significance:

> Kemp heard the same thing that I did, and I knew then that I was absolutely right in my anticipation. Electric waves which had been sent out from Poldhu had traversed the Atlantic, serenely ignoring the curvature of the earth which so many doubters considered would be a fatal obstacle. I knew then that the day would come on which I should be able to send full messages without wires or cables across the Atlantic...Distance had been overcome, and further developments of the sending and receiving apparatus were all that was required. (Tarrant, 56)[1]

A single letter that transformed communication. Ships were to be soon equipped with the technology, and most dramatically and poignantly of all, it was this medium through which the stricken *Titanic*'s predicament was communicated to rescuers and the world in April 1912. The most famous piece of Morse code in the world is *SOS*, *S* standing of course for 'save' and 'souls.' If ever proof of the eloquence of a single tiny sound were sought, Marconi's experiment that day on the edge of Newfoundland would surely provide it. Yet all great men stand on the shoulders of previous giants; Heinrich Hertz (1857–1894), John Ambrose Fleming (1849–1945), James Clerk Maxwell (1831–1879) and Michael Faraday (1791–1868) were all pioneers, whose work, reputations and contributions to the history of sound as an electrically transmittable medium relied on revelatory moments honed by endless and painstaking research and practical application. Of all these, the figure most directly linked in the chain of discovery to Marconi's first experiments was Samuel F.B. Morse (1791–1872). Morse, from Massachusetts, was a painter, and the story of his invention begins in tragedy. In 1825, working on a painting in Washington, Morse received a message by messenger to say that his wife was ill. The following day came a letter to say that she had died. Leaving Washington immediately, by the time he arrived home, the funeral had already taken place. Heartbroken, the incident caused an abrupt shift in career, and he decided to seek a means of instant communication. After discussions with Charles Thomas Jackson, an expert of the time in electromagnetism, Morse arrived at the concept of the single-wire telegraph.

Yet his devising of the code is itself a part of our story, insofar as it demonstrates the significance of the moment. Morse was determined to break the process down to the smallest units possible, seeking to create a code that contained representations of the most common letters through the shortest codes. Visiting the typesetters of his local newspaper, he observed how the printers organised their units, in cases, where each letter is within a separate compartment. There were more copies of the most common letters, because of course the printers needed more of certain letters than others in the process of setting a page of print. Thus, for example, Morse noticed that there were more 'e's than any other letter. This became his guiding principle, and he gave the letter 'e' the shortest code of all: a single dot. For Marconi's purpose in his great experiment, a single dot would have been too easily missed and could have been considered to be an accident of atmospherics. On the other hand, 's,' three dots... The world grows thus, from moments of discovery and decision, and the sonic evidence of these is all the more profound, because it is the most human of all.

Three small pulses travelled through air and transformed it forever, so we might ask the question, based on this, as to whether we affect a seemingly empty space by our very presence in it, even as we seek to record the experience of that space, either electronically or imaginatively. A room full of listeners changes the character of that room. A single person recording an unpopulated landscape negates the idea of human absence. By being present to hear, by being a witness to the see-able and audible, our consciousness contributes actively to the experience. We are makers at the same time as we are observers. Just as an apparently mute image does not offer silence but a receptacle in which the imagination can create its own personalised sound world, just as in darkness the mind may generate its own pictures to people the void, so may the implication of a small event call forth a mental response from us beyond the visual and cognitive. Be it conscious or subliminal, an internally articulated thought, it offers the potential of an imaginative signal offering interactive emotion. We may have linguistic limitations, but to move beyond words is to enter a world of sound suggested by the other senses that truly knows no frontiers. On Signal Hill, Newfoundland, a single letter of the alphabet opened a floodgate of global communication and said 'yes' to everything that subsequently followed it through the air.

GHOSTS IN A LANDSCAPE

The BBC producer Julian May has collaborated with the poet Katrina Porteous on a number of radio features that explore the thresholds between words and sounds conveying the narrative of place. One such programme made in 2007/2008 was *The Refuge Box*, a radio poem created for BBC Radio 3. The subject was a small cabin, built on stilts between Holy Island and the mainland of Northumbria, and it was built to save people cut off by the tide. (Holy Island becomes an island only at high tide, being linked to the shore by a causeway at other times.) The programme featured the voices of fishermen, coastguards and lifeboat men and others, as well as natural sounds of the grasses, winds and wild creatures of the terrain. All this, May blended with the poet's words, to create a piece in which all the elements paint a sound picture as virtually equal partners as they interact with one another, triggering meanings and elusive suggestions in the mind:

> Katrina spent some time recording not just people on the island, but sounds, and I recorded a lot, too – the wind in dry grass, the waves on the foreshore, birds. One amazing sound was the seals. They sound as if they are singing. When we listened it seemed they were questioning,

"Who, who?" Katrina worked this into her poem, a refrain questioning who the people were who had drowned over the years getting to or from the island. I think it is powerful, the way sound, language and meaning mingle.[2]

Porteous wrote of this moment in her essay, *A Rose with Many Centres*:

> From a distance, seals sound ghostly, a long drawn-out mournful 'Ooooo!' which seems both human and otherworldly. The ambiguities of this sound fascinated me...it seemed the aural equivalent of the reflection of the blue sky in the wet sand, the uncertainty of where land and water begin and end, the edge between what is human and what is not...In particular, the words 'blue', 'human', and the repeated question 'Who?' resonate within it, creating a kind of rhyme between the human music of speech and the inhuman, disembodied sounds of the place. (Porteous in Ivory and Szirtes, 114)

It is something the great wildlife sound recordist Ludwig Koch had noted many years before, while recording the sounds of seals in a Scottish cave, imagining the ghosts of drowned sailors in the unearthly songs; when the source of the sound is not itself visible, the meaning becomes negotiable. Many of Koch's recordings of birdsong however, while having an almost clinical precision in their execution, nevertheless sometime isolate the subject to the extend that the environmental context is altogether missing from the tape. What we hear is a bird singing, as if in an acoustically perfect studio, devoid of habitat, and thus to an extent meaning and imaginative engagement for the listener. When May and Porteous collaborated on location, recordings were made using professional facilities and with less sophisticated equipment, by the poet herself. Porteous remembered that:

> Interestingly, my rough, amateur field recordings of the seals worked better for this purpose than a sound recordist's much more detailed, close-up, specialist recordings. Perhaps that was because my recordings more faithfully captured the experience of hearing the seals from afar, mingled with the wind, across the sands.[3]

We possess the capacity to personalise the world internally, but it is an acquired skill, to train eye and ear to relate to one another, utilising memory to provide references. We may focus our attention on a voice, a song, a sound or a distant cry, but as we listen, the world goes on around us, and it is as if a choir is singing all the time, and one voice steps forward for a moment. The most efficient sound studios and anechoic chambers immure us from the sound world outside, but there are situations where we cannot but hear the sound under the sound, and it may define the moment beyond all else. When Julian May was recording *The Inexhaustible Adventure of a Gravelled Yard*, a location-based feature on the poet Patrick Kavanagh for BBC Radio 3 in 2018, one such moment gave him particular pleasure:

> One of the first things we recorded was the end of the programme, by his grave. At the end of presenter Theo Dorgan's piece I kept recording so I had some atmosphere to work with. We went on with our recording schedule, and over the next few days heard from many people as to how Kavanagh was a complex figure, rough and rude. He had a rather croaky voice.
>
> But he liked, got on with and was kind to children. When I came to put the programme together I realised that in the recording at the grave, just after Theo stopped talking, a crow cawed a couple of times, as if in approval, and in the background there was the sound of children playing on the playground of the village school. Sound chiming with what had been said, summing up the programme at the end. Maybe no one noticed, but I made sure we had a bit of this to fade out on.[4]

*

The world is where the world happens. Creative, social and sporting spaces such as stadia, concert halls, recording studios and theatres—even pubs, bars and clubs—have a strange, unexpected power when they are empty and as it were, inert, awaiting their next event while still reverberating from their last. Their purpose is to be occupied, and when they are dormant, they possess to the imagination a curious potency, a blend of anticipation and emotional vibration from the past. Small wonder that theatres are thought often to be haunted, for it is here that the intense immediacy of human creativity and interaction has shared experience, in some cases over hundreds of years. To stand in an acoustic space and actively listen can be an unnerving experience: a unique stillness presses on the ears, and it is as though the place itself is listening too. Where does all that sound go? If there could be such a thing as Place Memory, surely sound would be the key?

It is March 1940, in the Studio Albert in Paris, and a sound recordist is placing all his attention on the great harpsichordist and pianist Wanda

Landowska who is making a commercial recording of Scarlatti sonatas for release on *La Voix de son Maitre*. Yet what the microphone is about to inadvertently capture is a dark moment of history in one of the most remarkable recordings in all of music. Landowska was a Polish-born harpsichordist whose performances, teaching, recordings and writings played a large role in reviving the popularity of that instrument in the early twentieth century. She was the first person to record J.S. Bach's *Goldberg Variations* on the harpsichord. She became a French citizen and established the *École de Musique Ancienne*, Paris in 1925: from 1927, her home in Saint-Leu became a centre for the performance and study of old music. There is something about the harpsichord that gives it a civilised, precise expression. It is to the modern ear full of imagined historical echoes of elegance and decorous drawing rooms. Each note is distinct, every sounding, exact.

Landowska's recording session has gone well; she has recently recorded a number of Scarlatti sonatas both at the Studio Albert, and at the Studio de la Grande Armée, also in Paris. Now she is playing the Sonata in D Major, L.206, Kk 490. The instrument responds to her sensitivity. The music is full of nuance, at the same time containing some abrupt heavy chords as Scarlatti exclaims and the instrument seems to shout out. Suddenly, at precisely two minutes and one second into the recording, we hear something strange, an extraneous sound coming from outside in the street. A loud booming sound. An explosion. It is the very first symptom of a pivot in the balance of civilisation; it is the beginning of the Nazi invasion of Paris. Bombs and artillery explode in the background but amazingly, Landowska continues to record the Scarlatti sonata without stopping.[5]

It is an instant of profound significance, recorded on disc; the end of a story, and the beginning of a changed life for Landowska, a naturalised French citizen of Jewish origin. She was to escape with her assistant and companion Denise Restout, leaving Saint-Leu in 1940, staying for a time in southern France and finally sailing from Lisbon to the USA. She arrived in New York on 7 December 1941. The house in Saint-Leu was looted, and her instruments and manuscripts stolen. She never played in Europe again. Wanda Landowska died in Lakeville, Connecticut in 1959, at the age of 80.

Such moments put art into the context of the happening of history, and vice versa. For sound recordists, radio engineers and producers, location recording can be difficult, but it can also yield unexpected treasures and moment of intense witness. While the mind of the programme-maker may be focussing on a sound in the foreground of their consciousness, the microphone has no such hierarchy of attention. Recording in Athens on Good Friday morning in 1993, the BBC features producer Piers Plowright was recording the bells of a nearby church. 'Only on listening back do I hear just how many bells there were in the recording, from all over the city at different perspectives. The microphone has heard more than we have.'[6]

The microphone attends to everything, whether or not we do. It is not only small sounds that can be lost to our ears, because of course while the component parts in a soundscape may be audible, relative to one another, our brain makes the choices based on habit and 'need-to-know', so we may be aware of cause without noticing effect. In 2015, the Canadian producer Chris Brookes was recording a street band in Grenada, Nicaragua, as it approached him and then passed down the road:

> It is the custom in that city, on holy days, birthdays, etc., to set off "bombas", i.e. homemade firecrackers. Only thing is, one never knows how much gunpowder is enough, or too much. Anyway, some mischief-maker had set some of these at the side of the street, unbeknownst to the band, or to me. When they suddenly went off right in their path, the band was obscured by smoke, stopped playing. Too much powder in the bangers I think. Eventually the musicians recovered and marched forward again as the smoke dissipated. It wasn't until listening back to the recording that I particularly noticed the car alarms that had been set off by the bang. I'd been focusing on the band, then the shock of the explosions took my attention away. I hardly noticed the alarms until listening later.[7]

At the editing session that followed, listening to the sequence objectively, Brookes heard the alarms as a crucial punctuation in the picture as it moved through time. A marching band provides a sense of place, a feeling of atmosphere in its vernacular voice, but Nicaragua is an unpredictable environment where there is a sense that almost anything can happen at any time. The explosions were not malicious, but they posed an immediate bewilderment and disorientation. The alarms are in a way, a kind of metaphor for that, voices from the inanimate that speak for the human:

> What they do, I think, is provide an unexpected measure of the force of the blasts. I've found that people laugh at the bangs, but then laugh more heartily at the whoop-whoop of the alarms - because, I think, they're

surprised by the "comment" on the event that those quieter sounds make. Perhaps they are the most important sounds in the whole sequence.[8]

A street band passing by can be a poignant metaphor; time passing us, moving through a place. In 'Fetes,' part of his *Nocturnes* suite, the French composer Claude Debussy gives us just such a moment, as music is heard in the distance, almost interrupting the flow of the prevailing melody. It builds, comes closer, reaches an almost Bacchanalian climax and then gradually dissolves into the night again. There is a feeling that the sound actually changes the place as it passes through it, taking its music on to new ears, leaving a memory. Piers Plowright, recording an interview in Preston with an old man about his garden, encountered just such a moment. As happened for Chris Brookes, it was an unscheduled interruption that would bring new meaning to his radio feature, through accident, the unexpected and a 'found' comment provided by circumstance:

> Distantly the local brass band started up, and marched past us. It was quite unplanned and unprepared for; we had the sense to stand still and let it pass. It became the end of the programme and took on a new meaning, not felt in the moment: the passing of time, the inevitability of aging, and the way music always means more than its sound.[9]

For a producer building a programme, assembling the sounds after recording is composition in the truest musical sense of the word, and it is only at this point that the importance of some of these peripheral murmurs may become clear; indeed it may be the smallest incidental sounds that define the whole work. In early 2018, the British independent producer, Alan Hall, was making a music documentary with the American singer-songwriter-violinist Andrew Bird. The programme focussed on his relationship with his home city of Chicago, a city which he feels is haunted. In the programme, Bird talked about this sense of the presence of ghosts, and indeed of being one himself. While working on the documentary, Hall set out to record 'wild track,' the ambience of the city, some of which he would use to create the sonic landscape to complement the speech content in the finished piece:

> In a long stretch of "atmos," I recorded while walking along Michigan Avenue, I heard something spectral and ambiguous amid the car horns,

buskers, bus pneumatics and hubbub of traffic: a muted squeal, which I logged as "ghostly bus brake". Little did I know until well into the edit how important (totemic, metaphorical) this 4-5 second scrap would become, woven into Bird's music and the sounds of the city.[10]

When the story is told purely through sound, the link between maker and listener becomes intimate and intense, so that every minute sound is significant in making or breaking the engagement with narrative. Shortly after Hurricane Katrina hit New Orleans, Alan Hall was there with a former soldier, William Thompson. Thompson had been studying music, when, joining the National Guard, he was deployed to Iraq. He documented his war experiences on a CD of musical sound collages. Then, on his return, he came to another site of devastation in New Orleans, during the late summer of 2005. Hall was interested in the sounds on Thompson's CD as a source from which to tell this man's story; but Thompson was a chain-smoker, and as he was interviewed, the sound of his Zippo lighter punctuated the conversation. 'Eventually I realised the importance of this mundane action,' Hall recalled. 'It was both nothing and everything. In the compilation of the feature, the sound of a struck match or a lighter came to convey much more than the information that Will was lighting another cigarette' (Hall in Biewen and Dilworth, 97). It was the sound of a man reflecting, thinking, a symbol of the fact that in spite of everything, he was alive. Such is the relationship radio has with its audience that a sound becomes important even when it is not consciously acknowledged: we hear much more than we think we do.

> The ever-open ear takes in everything and, miraculously, seems to catalogue the continuous absorption of everyday – and everynight – sound in such a way that each association and resonance can be released, like Proust's madeleine moment, by the right trigger, the right key. (ibid., 98)

The term 'speech radio' is a misnomer, because speech is only one part of the orchestration that communicates meaning. We are part of a vast composition, in which great and small instruments juxtapose to engage with memory and imagination. A small sound dropped into a big place becomes itself big. A stone thrown into a river may sink, and the river does not stop, but for a moment, the sound of it has changed, and because we were there to hear it, we have shared a moment of experience that shifts into memory. At the start of his two-volume exploration of the Isle of Aran, Tim Robinson suggests that perhaps there was a moment when Time itself began. This leads him to an even bigger thought: 'If it is true that Time began, it is clear that nothing else has begun since, that every apparent origin is a stage in an elder process' (Robinson, 5). Sound, because it is temporal, is a part of this continuum, and because we too are only momentary visitors here, we respond to the smallest symptoms of existence. At least, we should; it is all there, waiting to speak, waiting only for us to focus our attention. In 1998, Hall and Piers Plowright collaborated on a radio feature about the great jazz pianist, Jimmy Yancey in Chicago. The programme took its theme from a melancholy piano blues called 'At the Window', a masterpiece of profound feeling. While preparing at the start of the trip, they were sitting in a hotel room setting up a recording. At this moment, it began to rain, and the sound tapped on the window of the room. Both men recalled the moment; in Plowright's words, 'Alan switched on the machine, and the sound of the rain on the window became a leit-motif in the documentary.' For Hall, 'that rain brought the window of 'at the window' audibly into being - and I seem to remember it was the very first thing we recorded.'[11]

KAFKA'S MIRROR

The natural world is full of such perspectives, but often they are at the edge of perception because they are not what we THINK matters most. Yet the tiny sounds are the movers and shakers that together make the soundscape, that cumulatively make the whole planet sound the way it does. The Natural History sound recordist Chris Watson finds himself consistently surprised by how uninvited sounds knock on the door of his equipment: 'Individual grains of sand striking my headphones in the Namib desert, fragmenting pressure ice in the Antarctic, pistol shrimps in tidal rock pools on the coast of Northumberland, creaking branches in the sacred forest around Mount Horaiji in Japan.'[12] To learn to value the natural world is to learn the value of *all* existence. Through the pace and hubbub of everything, it teaches us to listen to the sound under the sound until we arrive at silence, then listen again. Piers Plowright, as a young man in 1958 on active army service in the Malayan jungle, found himself a member of an ambush party, awaiting an enemy patrol: 'The stillness was intense. And then, ears sharpened by silence, I heard a tiny jungle stream, hundreds of yards away. Oddly calming, even at that moment.'[13] In the event, the patrol never came and the ambush never happened, yet the stream flows on, even now.

We learn to tune to such music, which is always there under the event, as we listen to the voice, but more than that, to the *sound* under the voice, the tone and the mood. Is there aggression there? Or fear? Is there a smile hidden, or a sob? We should listen actively and tell ourselves when our senses are becoming passive. That small voice, that whisper or tiny sound, that minute dot in the high, far sky is so easily missed, but we need to attend to it—it might be a matter of life or death. As J.A. Baker reminds us in *The Peregrine*, 'distance moves through the dim lines of the inland elms, and comes closer, and gathers behind the darkness of the hawk.'

Out of the darkness comes the whispering of memory, the smallest sounds that carry the most meaning of all. These may be, as for The Belgian producer Edwin Brys, the sounds of childhood, of lost voices from the past:

> Like being ill, lying on the couch in the living room, feverish. And hearing the other members of the family having dinner in the kitchen. Hearing their talk as through a gauze, a far distant murmur, nearly inaudible, but oh so recognisable, familiar and comforting. That sound formed the distance between them and me, they being in another room, able to eat and chat. Me being alone. These sounds created a warm solitude. Separated from but still embedded in the family; truly sounds at the edge of perception.[14]

For Brys, working to make stories out of audio, it is the sound asking a question, requiring attention, that captures the imagination through its ambiguity; this is what animates narrative:

> What is the most intriguing sound you ever heard? Not a deafening sound. Nor a sound so astonishing that you held your breath. Nor a sound so frightening that it made your flesh creep. No. It is three little knocks on a door. *Tap, tap, tap.* The most mysterious sound, the most engaging sound asks a question, opens on to a development, a reply, an action, then on again to another question…It is a sound compelling you to go forward.[15]

Our lives are framed in sound, and throughout time we learn of the world by asking such questions of it, provoked by a desire to understand what we hear. A child learns language by listening, and before that, learns that a sound will produce a response. Time passes. An ailing man or woman lies on a bed, emerges from a deep sleep, stirred by the song of a bird and asks 'where am I?' Time answers with sound: a comforting voice or a song. The smallest sound is enough to let us know that we exist. Between 1917 and 1919, Franz Kafka stopped writing entries in his diary, but continued to commit jottings of thoughts and ideas in a series of octavo-sized notebooks. These have become known as *The Blue Octavo Notebooks*, in their published form, and they begin and end with intimate sound: 'If someone walks fast and one pricks up one's ears and listens, say in the night, when everything around is quiet, one hears, for example, the rattling of a mirror not quite firmly fastened to the wall' (Kafka, 1). The final entry in the notebooks finds the traveller—whether it be Kafka himself or another, perhaps ourselves?—setting off on a long journey through 'barren fields, a barren plain, behind mists the pallid green of the moon...A horse is waiting, a servant is holding the stirrup, the ride takes him through an echoing wilderness' (ibid., 85). What are the sounds that echo through this wilderness? The horse's hooves? Its breathing? Our own thoughts? Whatever the sounds, hearing them proves that we are present in the world. We must continue to listen for Kafka's mirror as it moves slightly on its wall. The great film sound designer, Walter Murch, responsible for the sound editing of such movies as *Apocalypse Now* and *the English Patient* among many more, was once asked about his technical skill at layering sound on film, to which he replied:

> Ideally for me, the perfect sound film has zero tracks. You try to get the audience to a point, somehow, where they can *imagine* the sound. They hear the sound in their minds, and it really isn't on the track at all. That's the ideal sound, the one that exists totally in the mind, because it's the most intimate. (Walter Murch and Frank Paine in Weiss and Belton, 359)

Susan Sontag has written that 'photography is an elegiac art, a twilight art. Most subjects are, just by virtue of being photographed, touched by pathos...All photographs are *memento mori*.... Precisely by slicing out this moment and freezing it, all photographs testify to time's relentless melt' (Sontag, 15). On the other hand, sound too is a metaphor for our mortality because it is always disappearing, as temporal as are we. The loudest sound only emphasises the silence that surrounds it, so in its bleakest incarnation, sound too is a *memento mori*. Taken this way, the tolling of a single bell which may be seen on one level as a bridge between the material world and that of the spirit and the imagination may also be heard as analogous for life vanishing gradually into death.

While it may be melancholy to look at the image of a deceased loved one, yet it is harder still to listen to a recording of their living voice, the sound moving through time, as it once did in the everyday world of the living person. These things are poignant precisely because they can unexpectedly reach out and touch our consciousness 'like the delayed rays of a star' (Barthes, 2000, 81).

Likewise a silent moment made suddenly sonic can be timeless, as in the unexpected song of a bird. 'Birds,' said Olivier Messiaen, 'are the opposite of Time; they represent our desire for light, for stars, for rainbows and for joyful vocalises.'[16] Mark Doty has written, 'what we are is attention, a quick physical presence in the world, a bright point of consciousness in a wide field from which we are not really separate' (Doty, 68). In a vast orchestra, there are sounds from the smallest, subtlest instruments of which we may be unaware. Yet the music of the whole would not be the same without them. We are edging away from a mainland of noise into a calm sea of stillness, broken only by the lapping of the smallest sounds. It is these tiny morsels of sonic light that teach us to attend to the world, teach us the art of active listening. Pause, stillness, silence, the various species of space into which we pour our imagination are receptacles for prompts from which ideas and memories take flight. To repeat the mantra: because we hear, we imagine, and because we imagine, we see, and through it all there is an unheard rhythm through it all to which our lives move, the implied heartbeat that lies under iambic meter, an echo of that first great chord of sound that came upon us when we were born, and the world rushed into our consciousness, the low note beyond conscious hearing, which continues to reverberate through our lives. Poets have frequently been our companions on this sonic journey, so let a poet conclude it; Emily Dickinson, writing in about 1882, imagined what her last listening experience might be:

> I heard a Fly buzz - when I died -
> The Stillness in the Room
> Was like the Stillness in the Air -
> Between the Heaves of Storm -
>
> The Eyes around - had wrung them dry -
> And Breaths were gathering firm
> For that last Onset - when the King
> Be witnessed - in the Room -

I willed my Keepsakes - Signed away
What portion of me be
Assignable - and then it was
There interposed a Fly -

With Blue - uncertain - stumbling Buzz -
Between the light - and me -
And then the Windows failed - and then
I could not see to see - (Dickinson, 223)

So the last word goes to the minuscule insect that interrupted John Donne's conversation with God in an earlier chapter. At the start and at the end—through a baby's first cry to a dying sigh—we move between portals of sound, and those very first and last faint signals are the most meaningful sounds of all, signifying a key turning in a door of entry and departure. The rest may be silence, but Emily's fly is already moving on, busy with establishing a small sonic manifestation of its presence, blending with all the other murmurings in the whispering world.

NOTES

1. In fact a number of sources have raised doubts as the whether Marconi and his team actually heard what they claim to have heard. See https://www.theguardian.com/education/2001/dec/11/highereducation.news/However, Tarrant describes a number of subsequent experiments at Signal Hill on the following days, that recounted confirmation of reception from the original transmission (Tarrant, 57/8).
2. May, Julian. Personal correspondence.
3. Porteous, Katrina. Personal correspondence.
4. May, Julian. Personal correspondence.
5. Landowska, Wanda. *The Well-Tempered Musician: The Complete European Recordings, 1928–1940,* CD 8, track 1. Paris: United Archives, NUA 3, 2007.
6. Plowright, Piers. Personal correspondence.
7. Brookes, Chris. Personal correspondence.
8. Ibid.
9. Plowright, Piers. Personal correspondence.
10. Hall, Alan. Personal correspondence.
11. Plowright, Piers and Hall, Alan. Personal correspondence.
12. Watson, Chris. Personal correspondence.

13. Plowright, Piers. Personal correspondence.
14. Brys, Edwin. Personal correspondence.
15. Ibid.
16. Messiaen, Olivier. Programme note, *Quatuor pour la Fin du Temps*. Berlin: Sony Classical 88985363102, 2017.

BIBLIOGRAPHY

Books

Anderson, Hans Christian, trans. Keigwin, R.P. *Hans Christian Andersen Fairy Tales*, vol. 1. Odense: Flensted, 1953.

Assmann, Aleida. *Cultural Memory and Western Civilzation*. New York: Cambridge University Press, 2013.

Augaitis, Daina, and Lander, Dan. *Radio Rethink: Art, Sound and Transmission*. Banff: Walter Phillips Gallery, 1994.

Bachelard, Gaston. *Air and Dreams: An Essay on the Imagination of Movement*. Dallas: Dallas Institute Publications, 2011.

Bajac, Quentin. *Being Modern: MoMA in Paris*. Paris and New York: Fondation Louis Vuitton and The Museum of Modern Art, 2017.

Baker, J.A. *The Peregrine*. London: William Collins, 2017.

Barthes, Roland. *Image Music Text*. London: Fontana Press, 1977.

———. *Camera Lucida: Reflections on Photography*. London: Vintage Books, 2000.

Belinfante, Sam, and Kohlmaier, Joseph. *The Listening Reader*. London: Cours de Poétique, 2016.

Berger, John. *Ways of Seeing*. London: Penguin, 1972.

———, ed. Dyer. *Understanding a Photograph*. London: Penguin Classics, 2013.

Biewen, John, and Dilworth, Alexa. *Reality Radio: Telling True Stories in Sound*. Durham: University of North Carolina, 2010.

Blake, William, ed. Bronowski, Jacob. *William Blake: A Selection of Poems and Letters*. London: Penguin Books, 1975.

© The Editor(s) (if applicable) and The Author(s) 2019   127
S. Street, *Sound at the Edge of Perception*, Palgrave Studies in Sound,
https://doi.org/10.1007/978-981-13-1613-5

Bloch, Ernst, trans. Palmer, Peter. *Essays on the Philosophy of Music*. Cambridge: Cambridge University Press, 1985.

Bresson, Robert. *Notes on the Cinematographer*. London: Quarter Books, 1986.

Byrne, David. *How Music Works*. Edinburgh: Canongate, 2012.

Calvino, Italo, trans. Goldstein, Ann. *The Baron in the Trees*. Boston: Mariner Books, 2017.

Cavarero, Adriana. *For More Than One Voice: Toward a Philosophy of Vocal Expression*. Stanford: Stanford University Press, 2005.

Chion, Michel. *Audio Vision: Sound on Screen*. New York: Columbia University Press, 1994.

Clare, John, ed. Tibble, J.W. *The Poems of John Clare*, vol. 2. London: Dent & Co., 1935.

Coleridge, Samuel Taylor. *The Works of Samuel Taylor Coleridge*. Ware: Wordsworth Editions, 1994.

Cornish, Vaughan. *Waves of Sand and Snow and the Eddies Which Make Them*. Chicago: The Open Court, 1913.

———. *The Beauties of Scenery*. London: Frederick Muller, 1946.

Cox, Trevor. *Sonic Wonderland: A Scientific Odyssey of Sound*. London: Vintage Books, 2015.

Darwin, Charles. *The Descent of Man and Selection in Relation to Sex*. Princeton: Princeton University Press, 1981.

———. *The Voyage of the Beagle*. London: Meridian Books, 1996.

Dickinson, Emily. *The Complete Poems of Emily Dickinson*. London: Faber and Faber, 1977.

Dillard, Annie. *Pilgrim at Tinker Creek*. London: Canterbury Press, 2011.

Dolar, Mladen. *A Voice and Nothing More*. Cambridge, MA: The MIT Press, 2006.

Doty, Mark. *Still Life with Oysters and Lemon*. Boston: Beacon Press, 2001.

Dunn, Douglas. *The Noise of a Fly*. London: Faber and Faber, 2017.

Durham Peters, John. *The Marvellous Clouds: Toward a Philosophy of Elemental Media*. Chicago: University of Chicago Press, 2015.

Eliot, T.S. *The Music of Poetry: The Third W.P. Ker Memorial Lecture Delivered in the University of Glasgow, 24 February, 1942*. Glasgow: Jackson, Son & Company, 1942.

Foster, Charles. *Being a Beast*. London: Profile Books, 2016.

Frost, Robert, ed. Thompson, Lawrance. *Selected Letters of Robert Frost*. New York: Holt, Rinehart and Winston, 1964.

Gardner, Helen. *The Art of T.S. Eliot*. London: Faber and Faber, 1979.

Glennie, Evelyn. *Good Vibrations*. London: Hutchinson, 1990.

Godwin, Fay. *Land*. London: Heinemann, 1985.

Gregory, Richard L. (ed.). *The Oxford Companion to the Mind*. Oxford: Oxford University Press, 2004.

Hafez et al., trans. Davis, Dick. *Faces of Love: Hafez and the Poets of Shiraz*. New York: Penguin Books, 2013.

Harris, Alexandra. *Weatherland: Writers and Artists Under English Skies*. London: Thames and Hudson, 2017.

Harvey, Jonathan. *Music and Inspiration*. London: Faber and Faber, 1999.

Haskell, David George. *The Songs of Trees: Stories from Nature's Great Connectors*. New York: Viking, 2017.

Hill, Peter (ed.). *The Messiaen Companion*. London: Faber and Faber, 2008.

Hirschfield, Jane. *Nine Gates: Entering the Mind of Poetry*. New York: HarperCollins, 1997.

Hoban, Russell. *The Moment Under the Moment*. London: Picador, 1993.

Hooker, Jeremy. *Under the Quarry Woods*. London: The Pottery Press, 2018.

Hopkins, Gerard Manley, ed. Mackenzie, Norman H. *The Poetical Works of Gerard Manley Hopkins*. Oxford: Oxford University Press, 1990.

Hull, John. *Touching the Rock*. London: SPCK, 2013.

Ivory, Helen, and Szirtes, George (eds.). *In Their Own Words: Contemporary Poets on Their Poetry*. Cromer: Salt Publishing, 2012.

Jefferies, Richard. *Field and Hedgerow: The Last Essays of Richard Jefferies*. London: Lutterworth Press, 1948.

Kafka, Franz, trans. Kaiser, Ernst, and Wilkins, Eithne. *The Blue Octavo Notebooks*. Cambridge: Exact Change, 1991.

Keats, John. *Poetical Works*. London: Oxford University Press, 1967.

Koestler, Arthur. *The Act of Creation*. London: Picador, 1969.

Kroodsma, Donald. *The Singing Life of Birds: The Art and Science of Listening to Birdsong*. New York: Houghton Mifflin, 2005.

Lanoye, Tom. *Speechless*. London: World Editions, 2009.

Lawrence, D.H. *Selected Poems*. London: Penguin Books, 1965.

Laxness, Halldór, trans. Thompson, J.A. *Independent People*. London: Vintage, 2008.

Le Guin, Ursula K. *Steering the Craft: A 21st-Century Guide to Sailing the Sea of a Story*. New York: Mariner Books, 2015.

Leigh-Fermor, Patrick. *A Time of Gifts*. London: John Murray, 2004.

Levi, Peter. *The Noise Made by Poems*. London: Anvil Press, 1977.

Levinas, Emmanuel, ed. Hands, Seán. *The Levinas Reader*. Oxford: Blackwells, 1998.

Levitin, Daniel. *This Is Your Brain on Music*. London: Atlantic Books, 2008.

Lewis, Cecil. *Broadcasting from Within*. London: George Newnes, 1924.

Lin, Sonny (ed.). *Underwater Acoustics*. Jersey City, NJ: Clanrye International, 2015.

Locke, John. *An Essay on Human Understanding*. Oxford: Oxford University Press, 2008.

Longfellow, Henry Wadsworth. *The Poetical Works of Henry Wadsworth Longfellow.* London: Ward Lock, 1910.

Macfarlane, Robert. *Landmarks.* London: Penguin Books, 2016.

MacKenrdick, Karmen. *The Matter of Voice: Sensual Soundings.* New York: Fordham University Press, 2016.

Marston, Edward. *By Meadow and Stream.* London: Sampson Low, Marston and Company, 1896.

Melly, George. *Scouse Mouse.* London: Weidenfeld & Nicolson, 1984.

Miller, Hugh. *The Voyage of the Betsey, or a Summer Holiday in the Hebrides with Rambles.* Edinburgh: NMS Publishing, 2003.

Morton, Eugene S., and Page, Jake. *Animal Talk: Science and the Voices of Nature.* New York: Random House, 1992.

Motion, Andrew. *Keats.* London: Faber and Faber, 1997.

Nancy, Jean-Luc. *Listening.* New York: Fordham University Press, 2007.

Oliveros, Pauline. *Deep Listening: A Composer's Sound Practice.* Lincoln, NE: iUniverse, 2005.

———. *Sounding the Margins: Collected Writings 1992–2009.* New York: Deep Listening Publications, 2010.

Ong, Walter. *The Presence of the Word: Some Prolegomena for Cultural and Religious History.* Minneapolis: University of Minnesota Press, 1967.

Pietrasiewicz, Tomasz. *Theatre of Memory by the NN Theatre.* Lublin: Osrodek Drama Polska and Osrodek Brama Grodzka, 2017.

Perloff, Marjorie, and Dworkin, Craig (eds.). *The Sound of Poetry/The Poetry of Sound.* Chicago: The University of Chicago Press, 2009.

Reich, Steve. *Writings on Music, 1965–2000.* Oxford: Oxford University Press, 2002.

Richards, I.A. *Principles of Literary Criticism.* London: Kegan Paul, 1924.

Ricks, Christopher. *The New Oxford Book of Victorian Verse.* Oxford: Oxford University Press, 2008.

Robinson, Tim. *Stones of Aran: Pilgrimage.* London: Faber and Faber, 2008.

Rodenburg, Patsy. *The Right to Speak: Working with the Voice.* London: Methuen, 1992.

Rothenberg, David. *Why Birds Sing.* London: Penguin Books, 2005.

Sacks, Oliver. *Musicophilia: Tales of Music and the Brain.* London: Picador, 2012.

Serres, Michel. *The Five Senses: A Philosophy of Mingled Bodies.* London: Bloomsbury, 2016.

Shelley, Percy Bysshe. *The Major Works.* Oxford: Oxford University Press, 2009.

Smith, Patti. *Collected Lyrics 1970–2015.* London: Bloomsbury, 2015.

Sontag, Susan. *On Photography.* London: Penguin, 1979.

Stenger, Susan. *Sound Strata of Coastal Northumberland.* Newcastle: AV Festival, 2014.

Strauss, Richard, ed. Schuh, W., trans. Lawrence, L.J. *Recollections and Reflections.* London: Boosey and Hawkes, 1953.

Street, Seán. *The Memory of Sound: Preserving the Sonic Past.* New York: Routledge, 2015.

———. *Sound Poetics: Interaction and Personal Identity.* Cham: Palgrave Macmillan, 2017.

Takemitsu, Tōru. *Confronting Silence: Selected Writings.* Berkeley: Fallen Leaf Press, 1995.

Tarrant, D.R. *Marconi's Miracle: The Wireless Bridging of the Atlantic.* St. John's, Newfoundland: Flanker Press, 2001.

Tey, Josephine. *The Singing Sands.* London: Heinemann Books/New Windmill Series, 1970.

Thomas, Edward, ed. Thomas, R. George. *The Collected Poems of Edward Thomas.* Oxford: Oxford University Press, 1981.

Thoreau, Henry David. *Walden.* London, UK: Penguin Random House, 2016.

Toop, David. *Sonic Boom: The Art of Sound.* London: Hayward Gallery, 2000.

Trower, Shelley. *Senses of Vibration: A History of the Pleasure and Pain of Sound.* New York: Continuum, 2012.

Truax, Barry. *Acoustic Communication.* Norwood, NJ: Ablex, 1984.

Upton, Clive, and Davies, Bethan L. (eds.). *Analysing 21st Century British English.* Abingdon: Routledge, 2013.

Voegelin, Salomé. *Listening to Noise and Silence: Towards a Philosophy of Sound Art.* New York: Continuum, 2010.

Wearing, Deborah. *Forever Today: A Memoir of Love and Amnesia.* London: Corgi Books, 2005.

Weiss, Elisabeth, and Belton, John (eds). *Film Sound: Theory and Practice.* New York: Columbia University Press, 1985.

White, Gilbert, ed. Jefferies, Richard. *The Natural History of Selborne.* London: Walter Scott Ltd., undated.

Wordsworth, William. *The Prelude: Growth of a Poet's Mind.* Oxford: Clarendon Press, 1959.

Recordings

Adams, John/ New York Philharmonic/Maazel, Lorin. *On the Transmigration of Souls.* New York: Nonesuch Records, Nonesuch 7559 79816-2, 2002.

Deutsch, Diana. *Phantom Words and Other Curiosities.* San Diego, La Jolla: Philomel Records, Philomel-002, 2003.

Harvey, Jonathan. *Bird Concerto with Piano Song.* London: NMC Records NMC D177, 2011.

Heaney, Joe (Seosamh Ó hÉanai). *The Road From Connemara.* London: Topic Records TSCD518D, 2000.

————. *Say a Song: Joe Heaney in the Pacific Northwest.* Seattle: Northwest Archives Recordings NWARCD 001, 1996.

Landowska, Wanda. *The Well-Tempered Musician: The Complete European Recordings, 1928–1940,* CD 8, track 1. Paris: United Archives, NUA 3, 2007.

Lucier, Alvin. *I Am Sitting in a Room.* New York: Lovely Music LCD 1013, 1990.

McLean, Priscilla, and Korte, Karl. *Music from the Sounds of the Earth: Electroacoustic Compositions Created Entirely from the Sound of the World We Live in.* Baton Rouge: Centaur Records, 2012.

Messiaen, Olivier, perf. Sherlaw Johnson, Robert:*Catalogue d'oiseaux.* London: Argo Records, 2BBA 1005-7. 1973.

Messiaen, Olivier, perf. Frost, Martin, Debargue, Lucas, Jansen, Janine and Thedeen, Torlief. *Quatuor pour la Fin du Temps,* Berlin: Sony Classical 88985363102, 2017.

Reich, Steve/Kronos Quartet. *WTC 9/11.* Antwerp: Megadisc Classics MDC 7877, 2016.

Silvestrov, Valentin. *Bagatellen und Serenaden.* Munich: ECM New Series ECM 1988, 2007.

Takemitsu, Tōru. *Spirit Garden: Orchestral Works.* Leeuwarden: Brilliant Classics 8188, 2006.

INDEX

© The Editor(s) (if applicable) and The Author(s) 2019
S. Street, *Sound at the Edge of Perception*, Palgrave Studies in Sound,
https://doi.org/10.1007/978-981-13-1613-5